VIKING

RULE BRITANNIA

TRADING ON THE BRITISH IMAGE

ROBERT OPIE

AUTHOR'S NOTE

The dating of the items of this book has been determined in a number of ways. It is
possible to date precisely the advertisements from magazines (although some may
have been reused over several years) and showcards or display cards which were
coded by the manufacturer or printer. In some cases there is a handwritten date or
post mark; in others the image relates to a particular event such as a coronation.
Where an exact year cannot be attributed, an approximate date has been given. In
some cases it is difficult to know whether to date according to the original style or
to a minor update (for example, the paper bag on page 140 combines an engraving
from around 1865 with a post-1900 telephone number). Often, too, the design of a
label or tin remained virtually unchanged for decades, making exact dating of
individual examples difficult.

ACKNOWLEDGEMENTS

The advertising and packaging material reproduced in this book now forms part of
the Robert Opie Collection, much of which is on permanent display in the Museum of
Advertising and Packaging, in the Albert Warehouse, Gloucester Docks, Gloucester
(tel. 0452 32309).
I am most grateful to those who have helped me find these items over the years; to
the manufacturers for their assistance; to the Royal Warrant Holders Association;
to Irena St John-Brooks for her help in the editing of this book; to my mother for her
encouragement and support, and for the use of the Westerfield Library. I also owe
much to my father who guided me through the philosophy of collecting and from
whose own works I have drawn strength.

VIKING

Penguin Books Ltd, Harmondsworth, Middlesex, England
Viking Penguin Inc., 40 West 23rd Street, New York, New York 10010, U.S.A.
Penguin Books Australia Ltd, Ringwood, Victoria, Australia
Penguin Books Canada Ltd, 2801 John Street, Markham, Ontario, Canada L3R 1B4
Penguin Books (N.Z.) Ltd, 182–190 Wairau Road, Auckland 10, New Zealand

First published 1985

Typeset in Century Old Style

Designed by Stephen Kent

Printed in Great Britain by Jolly and Barber Ltd, Rugby

ISBN 0-670-80713-3

British Library Cataloguing in Publication Data available.

RULE BRITANNIA

TRADING ON THE BRITISH IMAGE

CONTENTS

By the time of Queen Victoria's Golden Jubilee in 1887, Britain was the most prosperous nation in the world. Her political, military and naval influence was at its height, the Empire encompassed a quarter of the world's population and, as a consequence of a successful industrial revolution, manufacturing output was still on the increase.

Showcard, *c.* 1887

Competition amongst manufacturers was intensifying. Thanks to an effective railway system, rival companies were able to send their wares further afield than before. No longer was it enough for a factory to cater only for its immediate locality. Survival of the fittest became the rule; and the ablest were gearing themselves to mass production, to new techniques of selling, and to supplying the growing markets of the industrial cities. Thus, of the hundreds of soap manufacturers operating in 1850, only a handful had survived by the turn of the century.

Yet towards the end of Victoria's reign Britain had to face the growing threat of competitive foreign goods. In order to repulse these new and unwelcome imports, Britain's manufacturers succeeded in rousing the nation to a mood of patriotism and pride by trading on the British image – which went down well not only in the home markets but across the Empire too.

During the 1870s rapid progress in the technology of reproduction and colour printing, especially offset lithography, meant that manufacturers could use pictorial advertisements more freely than ever before. Magazines of the 1880s were transformed by the increasing space given to illustrated advertisements. The end walls of buildings were enlivened by attractive pictures, rather than the repetitive verbiage of earlier times. The shops were now being furnished with highly colourful display cards to accompany the pyramids of equally colourful

packets mounted in the windows and on the shelves. Furthermore, this nation of shopkeepers was being well supplied with quantities of giveaway leaflets and paper advertising novelties, which often found their way into the scrap albums of children.

Detail from showcard, *c.* 1880

Leaflet, *c.* 1895

The innovations of the Victorian era, when coupled with the imports from the Empire, helped to create a wide range of products which were priced low enough to tempt the working man and his wife. Cocoa, tea, biscuits, chocolate, jams, condensed milk and meat extracts were no longer luxury goods. Soaps, scouring powder, dyes, starch, black lead, and boot and metal polish all were more effective and labour-saving.

'Though Borax Dry Soap is good enough for the highest in the land, it is cheap enough for the poorest, and is used in thousands of homes all over the Kingdom.'

With the increasing number of brands available, every

manufacturer knew that each one of his products had to be 'sold'. Descriptions of the brand were not necessarily enough and, when seemingly similar products were ranged side by side on the shelf, the one with the best image or most encouraging slogan would win.

Toothpaste pot, *c.* 1895   Feeding bottle and box, *c.* 1900

Brand names were obviously important and, while many were fanciful, sentimental or purely descriptive, a number caught the public's imagination by association with established figures, popular heroes or London landmarks. The Alexandra feeding bottle and Alexandra dentifrice, for instance, had a certain refinement. Heroes (and heroines) lent glamour to Baden-Powell Scotch whisky, Kitchener stove polish, Captain Webb matches, Stanley boot laces and the Grace Darling pen. London's famous landmarks were celebrated in products like Tower Bridge toffee, St Paul's brand cherry toothpaste and Carr's Big Ben biscuits.

Magazine insert, *c.* 1905

Soap magazine insert, *c.* 1900
Pen nib box, *c.* 1890
Chocolate box, *c.* 1910
See also the showcard on page 27 depicting the Grace Darling boot

Matchbox, *c.* 1895 (front and back)   Stove polish, *c.* 1910

The figure of John Bull was used many times, most memorably in the John Bull printing outfit, which was essential for every child, and the John Bull repair outfit, which was essential for every cyclist (though not if you rode with John Bull tyres, when you would have 'practically no use for this Outfit').

Unfortunately, foreign competitors also tended to brand their goods with patriotic British names, so that they would sell well in Britain: the Britannia box of slate pencils, for example, was made in Germany. In 1911 *Punch* warned, 'Patriotic purchasers are cautioned against buying goods of any sort marked "BRITISCH MADE"'. Many images offered opportunities for accompanying puns or witticisms. Ripolin paints, whose advertisement showed a soldier bearing the Union Jack, had the slogan 'The Colours that Stand'; Horlicks malted milk, whose standard-bearer was a girl, was 'Always up to the Standard'. British justice frequently pronounced favourable verdicts: 'Good judges drink only John Brown Scotch Whiskies', and 'Pears' Soap is used by all the best judges.' Golfing terms proved useful with 'Your tee is ready, Sir'–'No thanks, I'll have a Guinness' or 'Always fill your caddy with CWS tea.' Good use was made of many other sports: 'Scores every time' (football); 'Worth hunting for' (fox hunting); 'A good catch' (fishing); and 'The hit of the century' (cricket).

The Barham Press managed to encapsulate their company's abilities under various headings, each of which could be aptly illustrated in British images: Progressive, Attention, Excellence, Speed, Service.

Magazine insert, *c.* 1910

Publicity stamps, *c.* 1930

Slate pencil box, *c.* 1910

THE ROYAL CONNECTION

Tradesmen, manufacturers and shop-keepers have for centuries aspired to link their business with the royal family. The first known royal charter was granted in 1155 by Henry II to the Weavers' Company; and Royal Warrants have been bestowed on favoured suppliers ever since, but particularly during the reign of Queen Victoria. By the end of her reign no fewer than 1,080 firms were entitled to use the royal coat of arms (today it is some 850), which was displayed above the shop doorway, emblazoned on trade circulars, and proudly incorporated in wrappings of the product. This royal connection was indeed a most sought-after asset. Rowland's, of macassar oil fame, devoted half of their trade sheet in the early 1830s to telling the world of their prestigious associations, being 'Under the august patronage of His Most Gracious Majesty, Their Royal Highnesses The Dukes of York and Sussex, The Royal Family, Their Imperial Majesties the Emperor and Empress of Russia, and the Nobility and Gentry'. Those unable to gain the royal favour could console themselves with the arms of the guild to which they belonged, or even with a sign of a vaguely heraldic nature, designed by themselves, which, if they were shop-keepers, could be hung outside their premises.

Advertisement, *c.* 1830

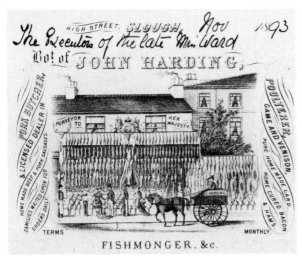

Bill head dated 1893

Book of trade advertisements, 1849

In a book of 1849 containing 562 advertisements for all manner of enterprises – manufacturers, banks, tradesmen, even spa resorts – no less than one quarter included the royal coat of arms. With the apparent glut of royal patronage, and the consequent lessening of its impact, it was not surprising that advertisements started to appear in the late 1860s which actually flaunted pictures of Queen Victoria and the Princess of Wales (later Queen Alexandra). Previously there had been isolated instances of the Queen's image being used on commercial wrappers and lids, sometimes alongside Prince Albert; but here were Queen Victoria, sipping Cadbury's cocoa in a railway carriage, and the Princess of Wales, wearing a flounced evening gown, apparently demonstrating the superiority of Glenfield starch.

Magazine insert, c. 1870

Magazine insert, 1871

Needle packet, c. 1840

Pin packet, c. 1850

Toothpaste lid, in use from 1870s until around 1920

Magazine advertisement, 1884

Commemorative tins for the Jubilee of 1897, Coronation of 1902
and subsequent royal visits of Edward VII and Alexandra:
mainly containers for chocolate, but also for butterscotch, tea and tobacco

Queen Victoria's Jubilees of 1887 and 1897, and Edward VII's Coronation in 1902, gave manufacturers of tea, biscuits and confectionery ample opportunity to issue commemorative, and highly decorative, tins, boxes and display cards, in which they were able to make the most of the recent advances in mass-production colour printing. These souvenirs were of course profusely ornamented with royal portraits and insignia. One way in which manufacturers were able to gain royal patronage was to thrust samples of their wares into the hands of royalty visiting the great trade fairs of the 1890s. The Prince of Wales was particularly susceptible to such gifts. While accompanying the Shah at the Paris Exhibition in 1889 he accepted a bottle of Bushmills whisky; in 1894, at Olympia, it was a casket of Terrabona tea and coffee, and in 1897 a sample of the new Maypole soap. Each event was recorded with alacrity in subsequent advertisements.

Postcard, 1902

By this time, advertisers were beginning to overstep the mark. The picture of Edward VII holding a cup of tea accompanied by the words 'Horniman's Pure Tea – A Right Royal Drink' was clearly going too far. By the 1920s most of these implied recommendations had been stopped, though the box for Cadbury's King George V chocolates still retained the picture of their royal patron, as did Rowntree's Prince of Wales chocolates. For royal occasions, the tradition of issuing souvenir tins and boxes remains.

Trade card, 1894.
This image was also used on their tea canisters

Chocolate boxes of the 1930s

Celebratory souvenirs of the Royal Wedding, 1981
Cadbury's Milk Tray, Mackintosh's Quality Street, Sharp's toffee,
Bassett's Liquorice Allsorts

SYMBOLS OF BRITAIN

For many centuries Britannia, the lion and John Bull have been the triumvirate representing the spirit of Great Britain. The image of Britannia was created by the Romans. A coin of Hadrian (*c.* AD 122) carries on its obverse the first personification of Britain: Britannia, with spear and shield, seated on a rock resembling the general outline of the island. (It is possible that the inspiration for Britannia was Minerva, the Roman goddess who was guardian of wisdom, art and commerce. Minerva was always depicted with helmet, spear and shield.) The more recent effigy of Britannia, on copper coins, dates from 1672; and the 50p piece of today continues the tradition.

Assurance companies were among the first to use Britannia as their symbol. London Assurance used Britannia in conjunction with the arms of London as their trademark from 1720, and many other fire assurance companies adopted the symbol during the nineteenth century.

London Assurance trade mark

Engravings from the policies of the Manchester, Phoenix and Caledonian Fire Insurance Companies

The association of lions with England dates back, appropriately enough, to the heraldic lion on Richard the Lionheart's shield. The lion has always been credited with the highest chivalric virtues, and was chosen as a badge by earls and knights, as well as by kings. The royal arms of Edward III were supported by a lion and an eagle from around 1350, and at this time also a lion was present in the royal arms of Scotland. Since 1603 the united royal arms of Britain have been supported by the familiar English lion and unicorn.

Needle packet, 1883 (front and back)

Salad oil bottle label, *c.* 1880

John Bull is first heard of as the national symbol of England in the seventeenth century; but it was Dr John Arbuthnot who established John Bull's reputation for bluff frankness and solidity in *The History of John Bull*, published in 1712 and reprinted in 1727 in *Miscellanies of Prose and Verse*. Although originally representing England, John Bull has often been the spokesman for the whole of Great Britain.

During the 1840s periodicals such as *Punch* (launched in 1841) helped to popularize Britannia, John Bull and, later, the British lion, by using them in political cartoons. By the 1880s they had become well-known and well-loved characters, and thus, when Britain needed to defend herself from the inflow of foreign goods, it was to these symbolic characters that British manufacturers turned when they needed to pursuade the British public to buy British.

Punch cartoons of the 1840s

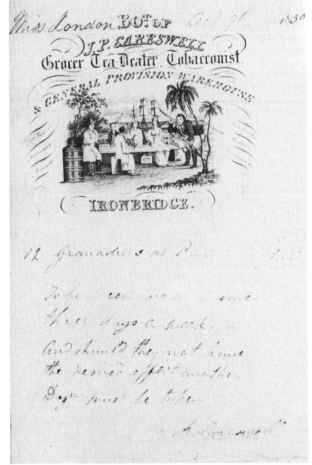

An engraved bill head dated 1830 in which a John Bull character appears saying, 'I hope I don't intrude'

Political poster, 1908

13

BUYING BRITISH

Towards the end of Queen Victoria's reign, British manufacturers could boldly announce that they were 'victorious over all', and advertisements depicted the nations of the world in agreement over such matters as the world's best sauce or strongest cocoa, Fry's being given this verdict by an 'international jury'. Other advertisements showed a variety of countries clamouring for such products as Cope's tobaccos or Page Woodcock's wind pills to cure indigestion. But as soon as other nations started to catch up, with their own industrial revolutions, 'agreement' was no more, and the rest of the world tried to make inroads on the British market. From the 1890s onwards, the manufacturers of those products under attack (anything from cigarettes and matches to mops and rubber heels) played upon the conscience of the would-be purchaser. 'Made by British labour with British capital', 'British made – British owned', 'Support British trade – buy home-made goods', 'These matches match our flag, beaten by none': these were all typical slogans in constant use around the turn of the century, often supported by Britannia or John Bull.

Magazine insert, *c.* 1890

Poster, *c.* 1895

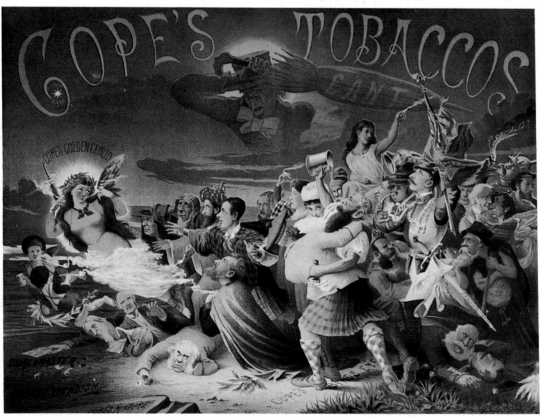

Poster, *c.* 1880

Leaflet, *c.* 1895

Magazine insert, *c.* 1890

Leaflet, *c.* 1895

Showcard, *c.* 1885

During the time when the American cigarette companies were attempting to persuade the British population to smoke their brands, Ogdens ceaselessly urged their supporters to remain loyal: 'Nail your colours [home industry] to the mast and have no parley with the foe of home industry, who offers an inferior substitute for Ogden's Guinea-Gold.'

Match manufacturers were also committed to the support of the home industry. Morelands of Gloucester, who encouraged patriots to 'Strike out foreign competition by buying England's Glory Matches', displayed advertisements showing a boxing ring in which John Bull lands a heavy punch on the nose of Germany. Another match company, R. Bell & Co., of Bromley-by-Bow, London, quizzed their customers on the outer wrapper of their Scottish Bluebell matches:

'About 50,000 gross of foreign made matches are imported weekly. Why? Because so few trouble to ask for, or instruct servants to ask for, English goods. These matches, boxes and labels are all made and printed in England. They are the very best quality and will not drop the ash. They are as cheap as the foreign. After this, which will you buy – English or foreign?'

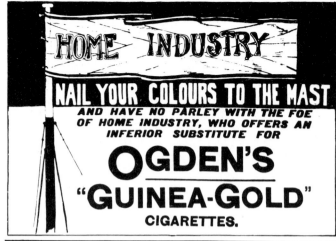

Magazine advertisements, 1900

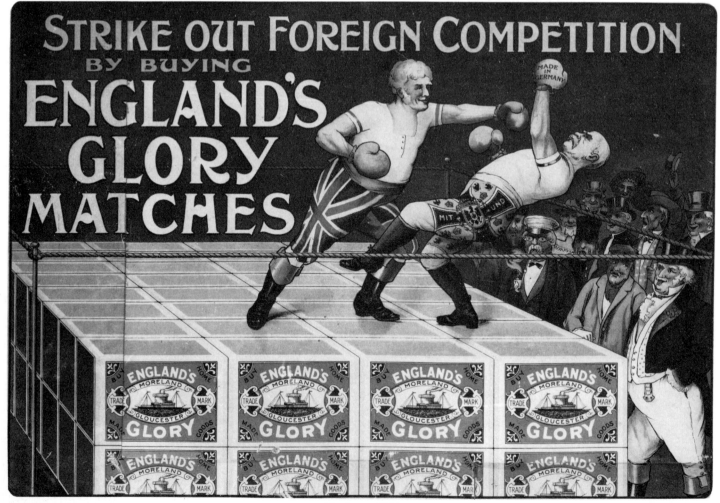

Tin sign, *c.* 1905

Trade press advertisement, 1911

DUDLEY & COMPANY,
Window Ticket and Poster Experts.

Detail from press advertisement of 1911, where the above company stated that the All-British Shopping Week, '...which is being organized in every town of any importance throughout the British Isles, will be the greatest Shopping Event of the Century, because it is a form of practical patriotism in which both the purchaser and seller can participate to their mutual advantage.
We have just prepared a Series of Special Tickets, Posters and Circulars with which to advertise these "All-British" Shows. They are smart, attractive and inexpensive; they strike the right keynote, and the spirit of "All-British" pervades them throughout. – Write for special illustrated list, No. D.R. 106, to-night.'

Newspaper advertisement, 1911

In 1911 an 'All-British Shopping Week' was organized for 27 March to 1 April. Under the slogan 'British Made – The Summit of Quality', the week was heralded as being the 'greatest shopping event of the century, because it is a form of practical patriotism in which both the purchaser and seller can participate to their mutual advantage'. History does not record, perhaps wisely, how successful this event was.

Matters must have reached a sorry state when it became necessary for a manufacturer of Union Jack flags to plead 'Why buy imported goods? Support your home industries.'

17

By the time of the First World War patriotic fervour was at its peak. Your country needed you, and the socks that patriots wore were made by Jason's: 'All-British in manufacture and in dependability are Jason's stockings and socks – be British in the selecting of all gifts . . . all made from the purest Australasian wool by British work people.' Boxes of patriotic Christmas cards would sell twice as fast as those of a more general nature. Readers of *Woman's Magazine* in 1916 were still being told 'Your country needs you' but this time it was 'to make sure that all you buy is British. Arden's Lustrous Crochet Cotton is British.'

By 1912, British Petroleum of Britannia House, London EC2, were able to claim they were 100 per cent British, and added an extract from a speech by the Prince of Wales at the inaugural dinner of the British Industries Fair: 'British goods are as good as any in the world and better than most.' Five years later Wakefield & Co., makers of Castrol, placed advertisements in the press announcing that Morris (the car manufacturers) said, 'Buy British and be proud of it, and approve Castrol, the product of an all-British firm.'

Buying British did not necessarily mean buying from Britain. Produce from the Empire was promoted in many ways, particularly since many of the materials for the goods bought in the shops came from countries within the Empire. Events like the Festival of Empire held at the Crystal Palace in 1911, and especially the British Empire Exhibition held at Wembley in 1924 and 1925, increased awareness of the Empire countries and their products. The Wembley Exhibition proved to be immensely popular, and such enthusiasm had not been generated since the Great Exhibition of 1851.

In 1926 the Empire Marketing Board was set up to promote home and Empire produce. The National Mark Scheme followed in 1929. Eggs, flour, fruit, vegetables and beef could all carry the National Mark. The scheme ensured that all who bought National Mark produce were getting 'the best our own countryside can supply. Produce bearing the National Mark is home-grown, graded and packed in the most up-to-date way.' The British Glasshouse Produce Marketing Association Limited issued its own promotion with 'British Tomatoes are best for health and flavour.'

Box of Christmas cards, *c.* 1915

Fruit can label, *c.* 1935

Poster issued by Ministry of Agriculture and Fisheries, *c.* 1935

Poster issued by Empire Marketing Board in 1928

Souvenirs and ephemera of the British Empire Exhibition held during 1924 and 1925

Throughout the country, individual shops organized their own Empire window displays, often featuring maps of the world with arrows showing where the produce had come from. Harrods had an 'Empire Week' in 1929, during which customers were invited to taste Empire Christmas pudding and Empire chocolates made from Empire cocoa nibs.

Although the 1930s saw less frequent use of John Bull and Britannia to advance the sale of British goods, patriotic slogans still abounded on packets and in advertisements. Fry's designed an emblem for some of their products, with the motto 'Is it British? Yes, it's Fry's.' Particular pride was taken in announcing that some brands were now home produced, such as Ryvita Crispbread, 'How Splendid! Ryvita is now British Made' (previously made in Sweden) and Welgar Shredded Wheat, 'Britons make it – it makes Britons' (previously made in the USA).

Cereal packet, c. 1935

Price list, c. 1931

Detail from advertisement, c. 1935

STIFF UPPER LIP

During the dark days of the Second World War, the British nation rallied to a single cause – one of survival; everyone had his part in Britain's 'big show', bringing out the best of the British character and its 'stiff upper lip'. Advertisements took up the message on a practical note (during the First World War the message had been mainly a patriotic one). Cutex nail polish was good for 'working hands' especially if you were a land-girl, and Bourn-vita ensured 'the essential body-and-mind restoring qualities from your sleep (even though it is interrupted)'. Kiwi claimed their boot polish gave the brightest blackout. Bovril gave strength to win.

Magazine advertisements, 1940

The post-war years saw continued rationing (butter, margarine and cheese being the last categories to be freed in 1954), and the countries of the Empire, who had contributed greatly during the war, now realized the attractions of independence and the Commonwealth.

The 1950s saw two morale-boosting events, the Festival of Britain (a celebration of the centenary of the Crystal Palace Exhibition in 1851, set up to show 'our pride in Britain's past, our confidence in her future') and in 1953 the Coronation of Elizabeth II, which combined nicely with the well-timed ascent of Everest, providing an atmosphere of celebration coupled with achievement.

Display card with packet, 1953

Poster, 1951

Leaflet and souvenir box, 1951

The arrival of commercial television in 1955 gave manufacturers a new and effective medium through which to promote their wares, and British images continued to be used on the small screen. Inevitably, much of the promotional budget now began to be channelled away from newspaper and magazine advertising. Furthermore, the arrival of the self-service stores and supermarkets towards the end of the 1950s hastened the decline of the showcard and the need for vast quantities of shop-window material.

British images still found their way into advertisements, however, though not as frequently or as boldly as before, reflecting the fact that Britain no longer held her former enviable trading position.

Souvenirs of the coronation of Queen Elizabeth II, 1953

21

Nevertheless, Britain continued to find an occasional use for a British lion, whether as a symbol of quality on eggs or as a mascot to rally English football supporters during the World Cup games of 1966. The endearing World Cup Willie character served as a pleasing association on selected products, notably Watney's ale, Lovell's confectionery and Willem cigars.

During the last twenty years the symbol of Britain most frequently used has been the Union Jack, almost to the exclusion of others, even Britannia and John Bull. In 1967 the 'I'm Backing Britain' campaign was launched and 'Buy British' themes continue today, the 'Think British Campaign' being the most recent. Other forms of flag-waving abound, with labels hanging from United Kingdom goods, and the federations for such produce as fruit, meat and cheese make extensive use of Union Jack symbols to encourage shoppers to buy British.

Showcard, 1959, with dummy eggs

1967 sticker

1980 sticker

1985 sticker

Eggs were stamped with the lion emblem from 1957 to 1968

Products and promotional material bearing the World Cup Willie character, 1966

Bread wrapper, 1985

Apple box and plastic bag, 1980

Plastic beer bottle and poster, 1984

Oil can, 1985

Boeing 767 model for Britannia Airways, 1984

British meat emblem, 1983

Poster, 1984

Foodmark quality symbols, 1985

Lard wrapper, 1985

NATIONAL PRIDE

Even though Britain now has a more cosmopolitan society than ever before and membership of the EEC is moulding her into litres and metres, and even though our fortunes balance precariously on events in the rest of the world, Britons, happily, remain steadfastly British (or, if preferred, English, Scottish, Welsh or Northern Irish). And today, products still trade on the images and symbols of Britain, although to a lesser extent. Manufacturers may prefer, however, to hark back to the days when Britain was great and glorious, when she beat the world, not only with the cricket bat but also with quality British products – backed up, it should be mentioned, with a small dose of gunboat diplomacy.

Indeed, manufacturers have often found inspiration in Britain's two-thousand-year-old heritage – sometimes the early conquerors, whether Romans or Vikings, sometimes heroes like Alfred the Great, Robin Hood and Sir Francis Drake; and then there were the valiant knights, and the Roundheads and Cavaliers from Cromwellian days.

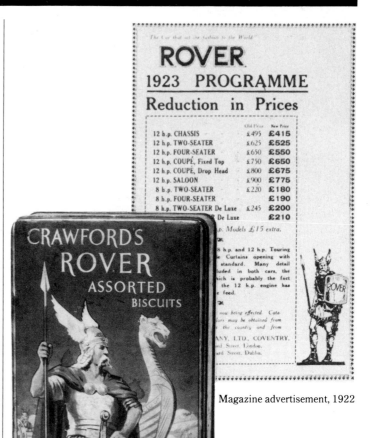

Magazine advertisement, 1922

Beer bottle label, *c.* 1925

Biscuit tin label, *c.* 1935

Biscuit tin , *c.* 1930

Showcard, *c.* 1900

Beer bottle, *c.* 1960

Showcard, *c.* 1920

Showcard, 1920

A poster of 1984 harking back
to the wartime spirit of the 1940s

Tobacco packet, *c.* 1910

Knife polish canister, 1890–1940

Toffee tin, *c.* 1925

Cake box, *c.* 1930

Cigarette packet, *c.* 1935

Milk cartons, 1984

Today Britain seems to have lost some of the impetus of earlier years; our former confidence has been sapped and we defend our shores from the scourge of continental milk by adding a Union Jack to the milk carton.

Surely now is the time for Britain to revive some of the spirit of the Victorian age, when the general sentiment was 'Victorious over all'; to rediscover the mood of 1915, when the manufacturers of Oxo cried 'Oxo is British, made in Britain, by a British company with British capital and British labour'; to regain the courage of a blitzed Britain when a Birmingham wine shop put up the notice 'We are carrying on with unbroken spirits.'

Undoubtedly Britain needs to find a new standing in the world, and this can be achieved through the quality, reliability, and competitiveness of her exported goods. Yet esteem for British industry has declined both at home and abroad. The first step towards recovering this esteem must be an understanding and appreciation of the part that her industry has always played, and must continue to play, for this nation to prosper.

Perhaps the images presented here from Britain's glorious past will give us inspiration for the future.

SYMBOLS OF BRITAIN

For many years John Bull and Britannia persuaded Britons to buy British, and the nations of the world to accept that British was best. But soon after the Great War of 1914–18 their appeal waned and, at this time also, manufacturers found themselves increasingly under attack from foreign competition.

The British lion featured in many advertisements, often with the Union Jack, though not all the lions, particularly those in trade marks, were necessarily characterized as British. Indeed, bulldogs were more often used for their tenacity than for their connection with John Bull. Bulldog clips are a case in point.

Bale label, *c.* 1885

Leaflet, *c.* 1900

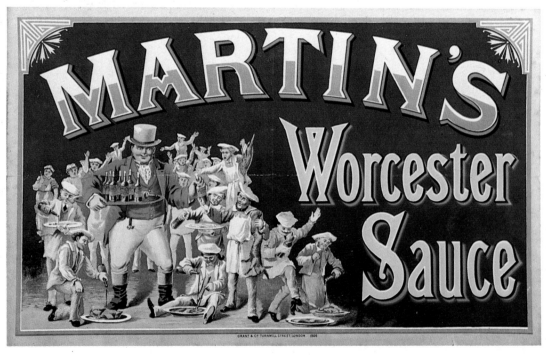

Poster, *c.* 1885

Showcard, *c.* 1895

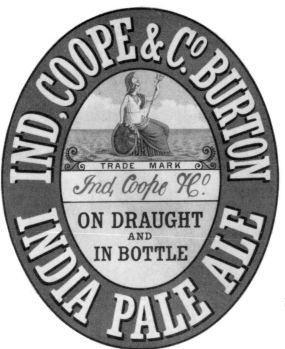

Display label, *c.* 1890

The Britannia trade mark was probably being used
by Ind, Coope from 1862, and was formally registered
by them in 1876, following the passing of the Trade
Marks Act in 1875 which provided protection from imitations

Paper folding novelty, *c.* 1890

Trade mark in use
during the 1890's

Leaflet, *c.* 1905

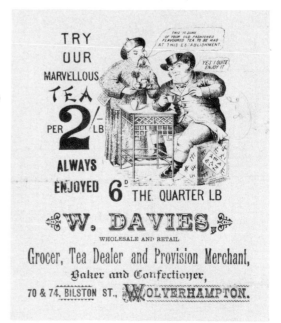

Leaflet dated 1887

Magazine advertisement, 1886

Showcard, *c.* 1910

Magazine advertisement, 1900

Patriotic slogans covered the back of these matchboxes of the early 1900s: "Nearly two million unemployed, yet half the matches used in England are made abroad, and this factory is on short time – buy British matches."

Magazine insert, c. 1900

Varnish tin, c. 1920

Magazine advertisement 1906

AN IDEAL FOOD
For INFANTS, CHILDREN, and INVALIDS.

Magazine advertisement 1896

Poster 1899

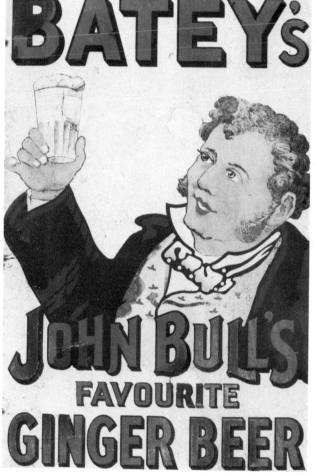

Enamel sign, c. 1900

29

Poster, *c.* 1905

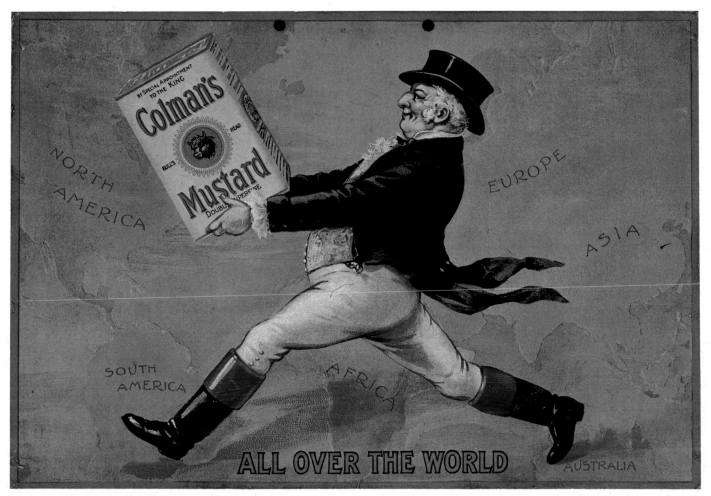

Showcard, *c.* 1905

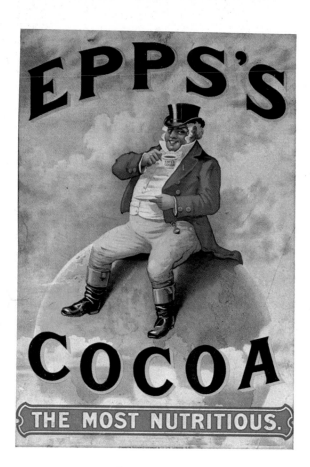

Showcard, 1899

Novelty card, *c.* 1900

Cut-out card, *c.* 1910

COLONIAL PREMIERS: "Well, good-bye, John, we have thoroughly enjoyed ourselves, We are proud of our Queen, and of our Empire. May the sun never set upon it."
JOHN BULL: "It's not likely to."
COLONIAL PREMIERS: "Never, so long as we get such a right royal SUNLIGHT welcome, SUNLIGHT weather, and plenty of —er——"
JOHN BULL (Smiling): "SUNLIGHT SOAP."

Magazine advertisement, 1897

Magazine advertisement, 1889

Magazine advertisement, 1900

Magazine advertisement, 1896

Magazine advertisement, 1898

Magazine advertisement, 1898

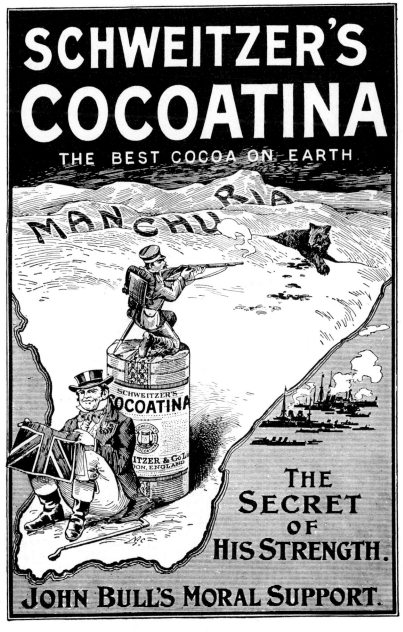

Magazine advertisement, 1904

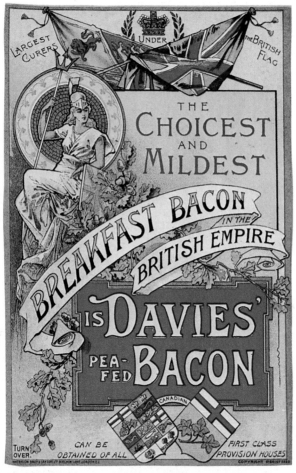

Leaflet, *c.* 1905

Showcard, *c.* 1910

Showcard, *c.* 1915

Sauce bottle, *c.* 1880

Tobacco tin, *c.* 1920

Playing card packs, *c.* 1910

34

Label for outer box dated 1904

Biscuit tin label, *c.* 1925

Trade card, *c.* 1905

Biscuit tin label, *c,* 1925

Tobacco tin label, *c.* 1925

Trade card, *c.* 1900

Trade card, *c.* 1895

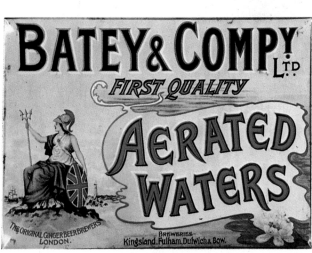

Showcard, *c.* 1910

Magazine insert, *c.* 1895

Leaflet, c. 1905

Magazine advertisement, 1917

Magazine advertisement, 1910

Magazine advertisement, 1900

Magazine insert, c. 1895

Blotter card, c. 1920

Showcard, c. 1895

Price list, 1932

Magazine advertisement, 1916

Magazine advertisement, 1916

"VOLKER" MANTLES

BRITISH
MADE

SUPPORT BRITISH TRADE

Poster, c. 1905

SINGER
JUNIOR · SENIOR · SIX

Britain's Finest
Car Value —

Magazine advertisement, 1928

Your Country
needs you

ARDERN'S
LUSTROUS CROCHET COTTON

Magazine advertisement, 1916

PIONEER
RANGE

Miniature advertising sticker, c. 1915

Matchbox, c. 1925

SHELL

"The British Lion yields the power Britannia wields"

Promotional booklet back cover, 1924

GREAT
FREE
OFFER!!

Serv-ol

Magazine advertisement, 1915

PLAYER'S
NAVY CUT

Stands on Quality at

Showcard, 1954

R. White's trade mark, c. 1900

Metal safe plate, c. 1890

CITY MEAT
DOG CAKES
WALKER HARRISON

Trade card, c. 1905

Carson's   *Empire* CHOCOLATES

Chocolate box, c. 1930

EMBLEMS OF THE SOVEREIGN

The height of respectability for a tradesman or a manufacturer was achieved when a royal warrant was granted by the sovereign or a member of the royal family. Naturally proud of royal recognition, firms would emblazon the coat of arms on labels, tins and pot lids, leaflets, posters and showcards, thus announcing by implication that the king or queen was partaking of this product.

The emblem of a crown was on occasion used to good effect, though this was not widely practised.

Display card, *c.* 1895

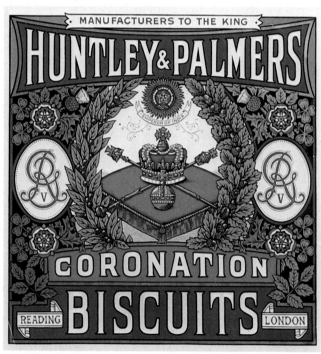

Biscuit tin labels issued for the Coronation of George V in 1911

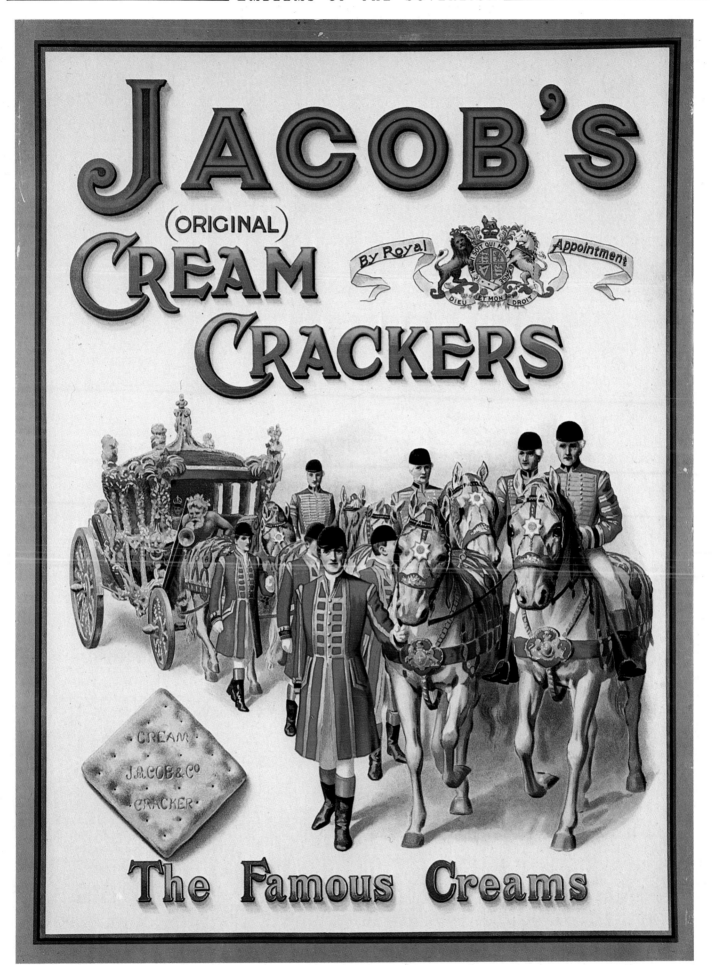

Showcard, 1911

Advertisement, *c.* 1850

Bill head dated 1847

Advertisement, *c.* 1840

Biscuit tin label, *c.* 1900

Soda water label, *c.* 1870

Bill head dated 1805

Anchovy paste pot, *c.* 1885

Matchbox, *c.* 1890

Towel label, *c.* 1910

Trade card, *c.* 1850

Trade card, *c.* 1830

Cake box labels, *c.* 1905

Button label, *c.* 1920

Showcard, *c.* 1890

40

Glass sign for shop window, *c.* 1905

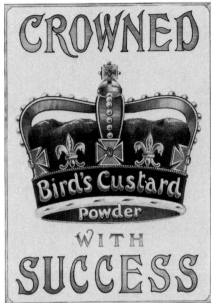

Magazine advertisement, 1902

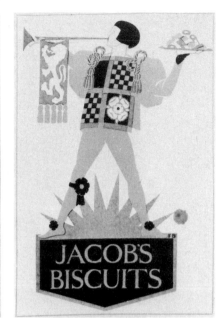

Showcard, 1928

Stout label, *c.* 1925

Soft drink label, *c.* 1930

Cigarette box, *c.* 1910

Soft drink label, *c.* 1930

Magazine advertisement, 1902

Magazine advertisement, 1898

Magazine advertisement, 1928

Glass sign for shop window, *c.* 1905

Showcard, c. 1910

Two-layer display card, c. 1925

Showcard, c. 1905

Showcard, 1897

42

THE ROYAL PRESENCE

The royal family have held a place in the nation akin to that of the gods on Olympus: far from the common people, but with human needs like theirs. Thus it was reassuring to know that the sovereign also drank tea or cocoa and seemingly 'recommended' a particular brand. Some manufacturers dared to go further by naming their products after royalty, such as Victoria window blind cord or Alexandra oil.

Today, images of the royal family continue to enter homes via television or magazines. Apart from the granting of royal warrants, however, the royal family does not endorse commercialization, with the exception of the commemoration of royal weddings, coronations or jubilees, when firms may receive permission to issue celebratory tins and packets.

Trade card, 1902

Trade card, 1897

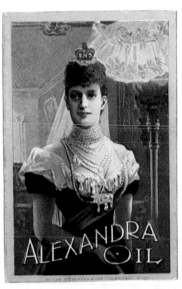

Trade card, *c.* 1905

Showcard, *c.* 1880

43

Bale cloth label, *c.* 1870

Magazine insert, *c.* 1870

Leaflet, 1893

Magazine advertisement, 1887

Leaflet, *c.* 1895

Trade card, 1896

Magazine advertisement, 1897

Pen nib box, *c.* 1895

Magazine advertisement, 1887

Window-blind cord packet, *c.* 1890

44

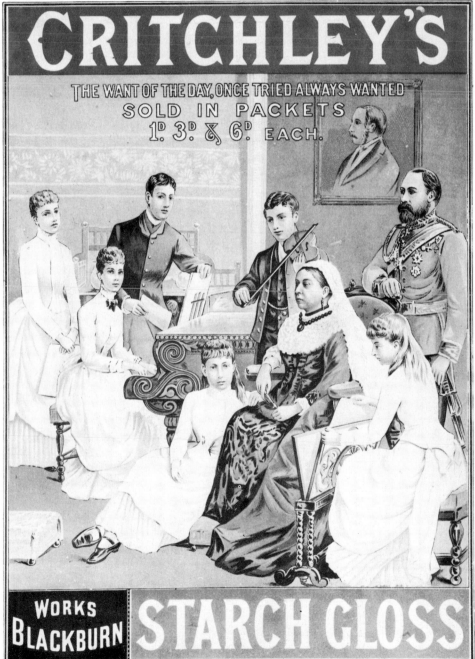

Showcard, *c.* 1885

Bale cloth labels, 1890s

Leaflet, *c.* 1890

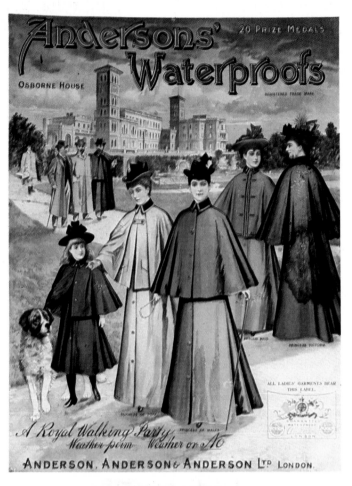

Showcard, *c.* 1900

Magazine insert, *c.* 1900

Magazine insert, *c.* 1900

Bookmark, 1902

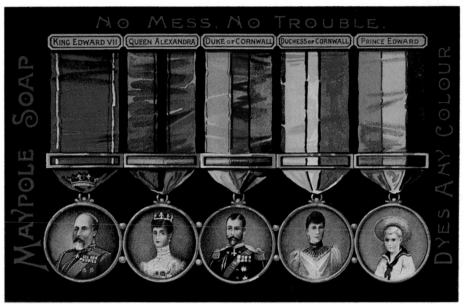

Magazine insert, 1905

Cigarette tin, *c.* 1900

Leaflet, 1897

Pomade pot label, *c.* 1890

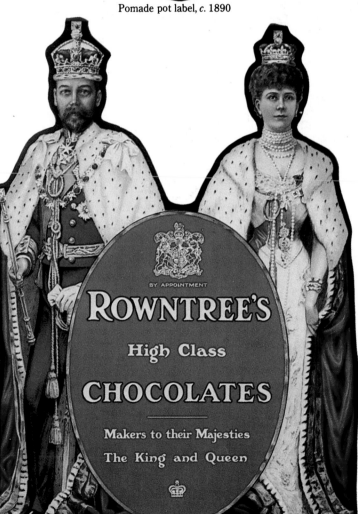

Display card, 1911

Matchbox label, 1902

Matchbox label, 1911

47

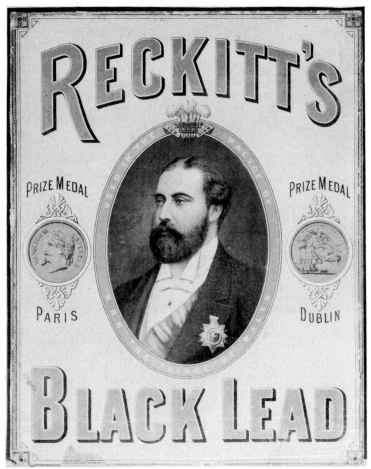

Showcard, *c.* 1890

Magazine advertisement, 1892

Magazine advertisement, 1905

Cigar box inner label, *c.* 1905

Matchbox labels, 1902

Trade card, 1902

Magazine advertisement, 1897

Leaflet, 1911

Magazine advertisement, 1897

48

Showcard, c. 1905

Matchbox label, 1902

Cigarette tin, c. 1905

Matchbox label, c. 1905

Showcard, 1911

Magazine insert, c. 1900

Trade card, c. 1900

HUNTLEY & PALMERS BISCUITS

Magazine advertisement, 1903

Advertisement, c. 1930

Showcard, c. 1930

The British flag is an emotive sign, arousing patriotism in Britons in a foreign land or in the thick of battle, and indicating the nationality of a boat or a manufacturer's brand or the whereabouts of the sovereign (indicated by the presence of the Royal Standard which, in its present form, dates back to 1837 when Queen Victoria came to the throne). The Union Jack unites the crosses of St George, St Andrew and St Patrick and has been flying in this form since 1800. No other image so quickly conveys the message that a product is made in Britain, even though today the flag is less often depicted flying from a flagpole.

Tobacco tin label, *c.* 1905

Advertising novelty, *c.* 1900

Shaped trade card, *c.* 1895

Biscuit tin label, *c.* 1905

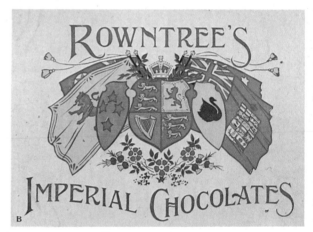

Chocolate box label, *c.* 1905

Cake box label, *c.* 1905

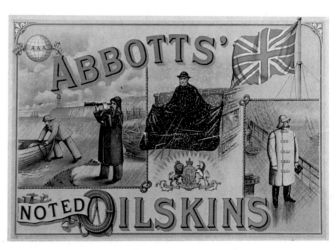

Showcard, *c.* 1895

Showcard, 1915

51

BOVRIL

BY ROYAL WARRANT TO THE KING.

Poster 1903

AT THE FRONT

LEMCO

The Product of 5,000 BULLOCKS Already Shipped to the BRITISH FORCES.

Magazine insert, *c.* 1900

ASPINALL'S ENAMEL

ASPINALL'S OXIDISED ENAMEL

The finest colors in the world.

BEWARE OF IMITATIONS.

Magazine advertisement 1897

HORLICK'S MALTED MILK

ALWAYS UP TO THE "STANDARD"

Magazine insert, *c.* 1910

"BOY SCOUT" (REGD) BISCUITS

MACKENZIE & MACKENZIE, EDINBURGH.

Showcard, *c.* 1910

Paper bag, *c.* 1915

Showcard, *c.* 1920

Paper flag, 1902

Biscuit tin label, *c.* 1905

Matchbox label, *c.* 1880

Illustration of window poster, *c.* 1910

Leaflet, 1902

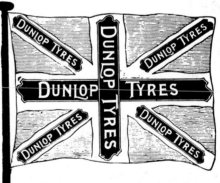

Magazine advertisement, 1902

Chocolate biscuit tin, *c.* 1890

Cigarette packet, *c.* 1910

Cigarette packet, *c.* 1930

Cigarette packet, *c.* 1920

Tobacco tin, *c.* 1920

Pin tin, *c.* 1900

Playing cards, *c.* 1920

Matchbox label, *c.* 1920

Razor blade packet,
c. 1920 (front and back)

Matchbox, *c.* 1910 (front and back)

Tacks box, *c.* 1915

Dentifrice powder tin, *c.* 1910

HUNTLEY & PALMERS
TRIBREK
THE BRITISH BREAKFAST FOOD

Three dimensional display card, *c.* 1935

Chocolate box, *c.* 1900

Beef cubes tin, *c.* 1915

Two-layer display card, c. 1935

Showcard, 1915

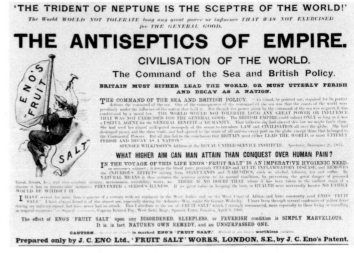

Beer bottle label, c. 1905

Showcard, c. 1910

Magazine advertisement, 1901

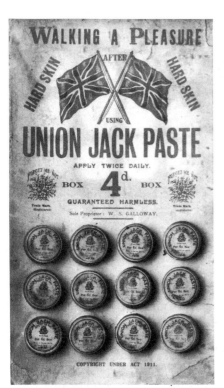

Product display card, c. 1915

c. 1910

Magazine advertisement, 1895

Ink bottle label, c. 1910

Whisky bottle label, c. 1910

Soap wrapper, c. 1910

56

The National Condiment

Promotional booklet (back cover), 1919

Mustard tin labels, *c.* 1920

Paper folding novelty, *c.* 1910, it builds up
the Union Jack from the flags of St George,
St Patrick and St Andrew

Springwell
Table Water

A TABLE WATER
OF UNSURPASSED QUALITY.

On Sale at all the Bars
in this Theatre.

BRITISH-*Every Drop!*

Theatre programme, 1916

Showcard, *c.* 1915

Soft drink label, *c.* 1925

Wine label, *c.* 1910

Soap box with overwrap which contained a free Union Jack flag, *c.* 1930

Stout label, *c.* 1930

RULE BRITANNIA

TRADING ON THE BRITISH IMAGE

BRITAIN'S FIGHTING FORCES

The immense pride that Britain had in her army and navy made them natural allies for advertisers. Particularly during the Boer War, heroes abounded and were portrayed on tins and packets as well as in advertisements. The new battleship class of Dreadnoughts, first launched in 1906, also launched a number of products; there were Dreadnought boot heels, Dreadnought cigarettes and even Dreadnowt razor blades. Military, navy and, later, airforce images had strong links with tobacco and cigarettes, since smoking was very much part of the services' way of life.

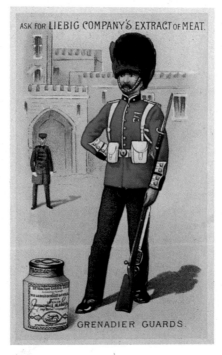

Series of six Liebig cards issued around 1890

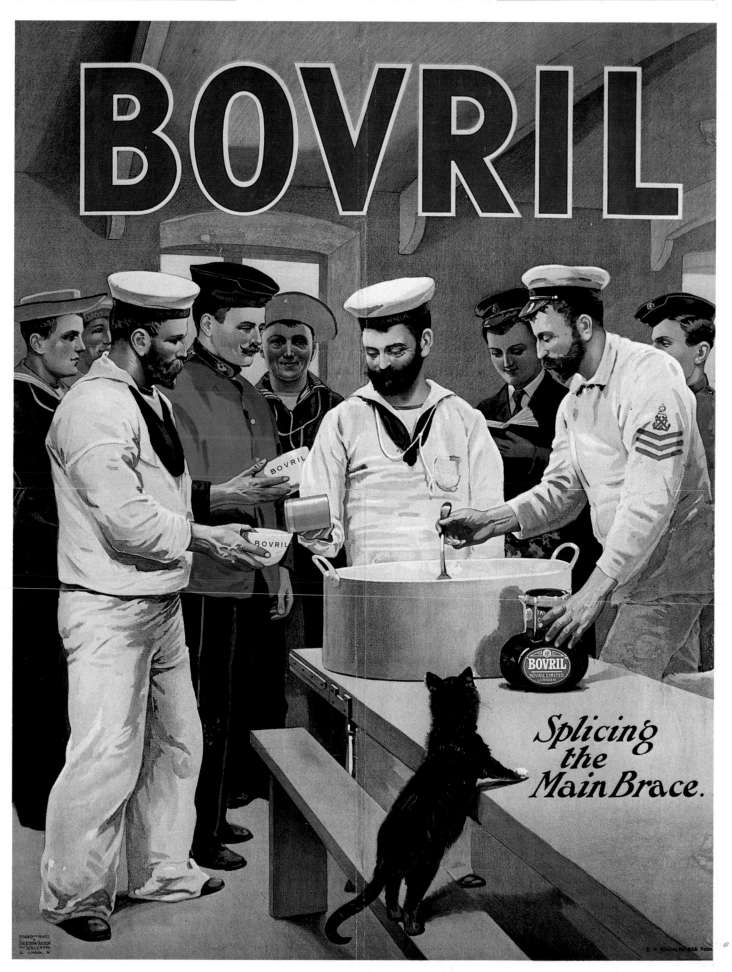

Showcard, 1903

Information cards, *c.* 1895

Magazine advertisement, 1898

Magazine advertisement, 1896

Matchbox label, c. 1890

Matchbox label, c. 1910

Matchbox label, c. 1900

Magazine advertisement, 1888

Whisky bottle label, c. 1900

Leaflet, c. 1905

Leaflet, c. 1900

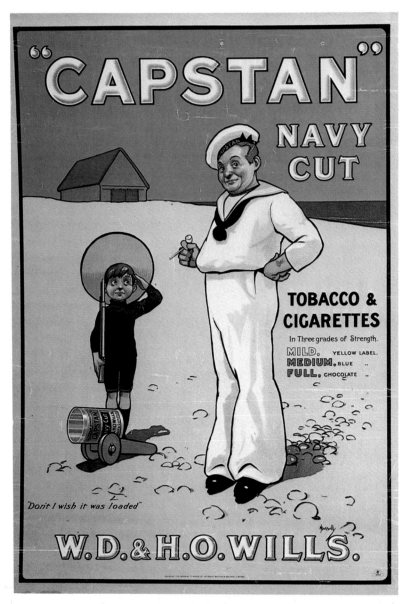

Showcard, 1907

Magazine insert, 1911

Shaped trade card, c. 1895

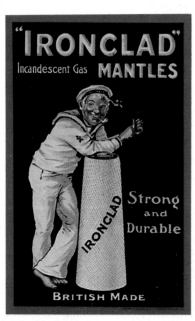

Postcard, c. 1910

Bookmarks, *c.* 1900

Postcard, *c.* 1905

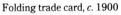

Folding trade card, *c.* 1900

Bookmark, *c.* 1900

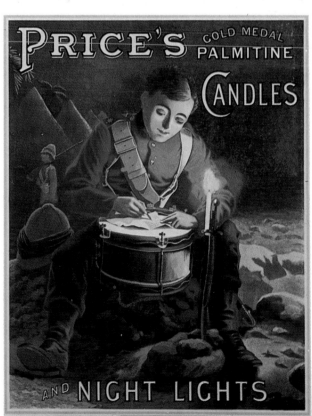

Showcard, *c.* 1900

Showcard, *c.* 1910

63

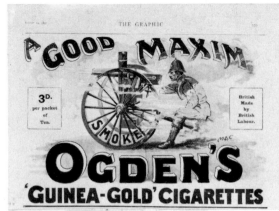

Magazine advertisement, 1897

Magazine advertisement, 1900

GEO. A. BLACKBURN,
18, NORTHGATE,
HALIFAX.
Given away with One Pound 2/- TEA.

Shopkeeper's gift calendar, 1899

Matchbox label, c. 1900

Cloth bale label, c. 1900

Leaflet, c. 1900

Pattisons' Whisky in GENERAL use

"IN GENERAL USE."

A Commanding Spirit finds its way to the front. PATTISONS' WHISKY commands success because it has been found by the public to be a genuine, wholesome, palatable beverage, carefully blended and thoroughly matured. It is cream-like in taste, with all the stimulating qualities of the pure Highland spirit. Sold Here, There, and Everywhere.

Sole Proprietors: PATTISONS, LTD., Highland Distillers, BALLINDALLOCH, LEITH, AND LONDON.

Magazine advertisement, 1897

Magazine advertisement, 1900

Showcard, c. 1910

Showcard detail, c. 1905

Magazine advertisement, 1899

Magazine advertisement 1900

Product display card, c. 1910

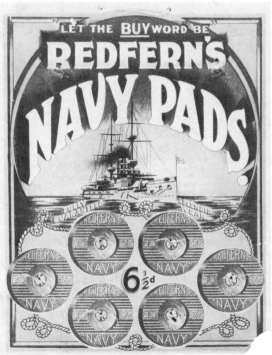

Product display card, c. 1910

Showcard, 1903

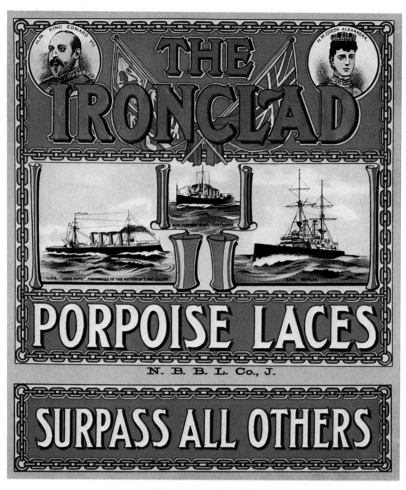

Lace box label, *c.* 1905

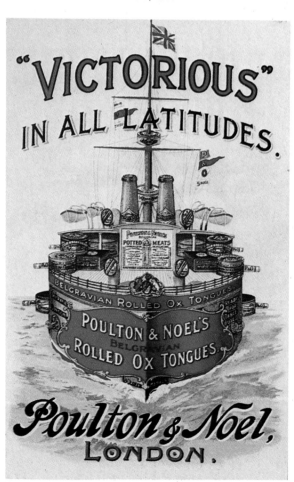

Magazine insert, *c.* 1895

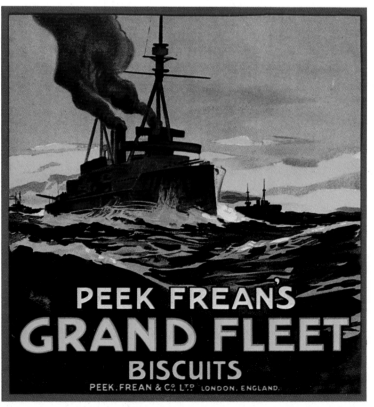

Biscuit tin label, *c.* 1920

Trade card, *c.* 1890

Card sample holder, *c.* 1895

Pin folder, *c.* 1900

Matchbox, *c.* 1920

Snap fastener card, *c.* 1915

Poster, *c.* 1895

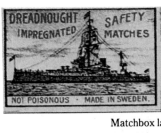

Matchbox labels, *c.* 1910

Showcard, 1929

Dye packet, *c.* 1940

Showcard, *c.* 1935

LIFEBUOY SOAP

JACK THE GERM KILLER.

"DON'T run away with the idea that we only thrive on the ozone that is blown at us," says Jack. "We can't get along without some of the good things of life aboard. What do you say, mates?"

Jack, you deserve the very best the nation can supply. From our own point of view All Sea and No Soap would make you a very dull boy, and we are delighted to have the privilege of supplying you with the soap which makes you happy and healthy.

Lifebuoy is an ideal soap for bath and toilet. It cleans and disinfects at the same time. It cleanses, invigorates and keeps the skin healthy. The mild carbolic odour you note in Lifebuoy Soap is the sign of its splendid protective qualities.

MORE THAN SOAP, YET COSTS NO MORE.

Send him a Tablet in his next parcel; he will appreciate it.

LEVER BROTHERS LIMITED, PORT SUNLIGHT.

Magazine advertisement, 1916

SUNLIGHT SOAP

£1000 SUNLIGHT

GUARANTEE OF PURITY

WHILE SUCH QUALITY EXISTS VICTORY IS ASSURED.

SUNLIGHT SOAP is victorious in the wash by land and by sea, because it represents the Highest Standard of Soap Quality and Efficiency.

In the air the triumph of Sunlight Soap is still more manifest, because clothes washed with Sunlight dry white as the driven snow. It is used and appreciated by the cleanest fighters in the world.

£1,000 GUARANTEE OF PURITY ON EVERY BAR.

Include a Tablet in your next parcel to the Fleet or Front.

The name Lever on Soap is a Guarantee of Purity and Excellence.

LEVER BROTHERS LIMITED, PORT SUNLIGHT.

Magazine advertisement, 1916

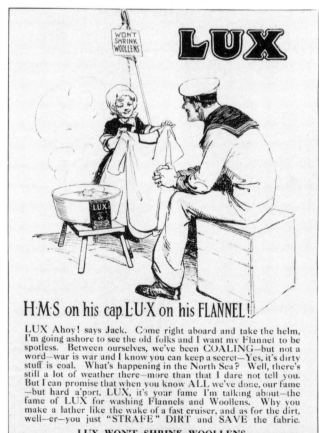

WON'T SHRINK WOOLLENS

LUX

H·M·S on his cap L·U·X on his FLANNEL!

LUX Ahoy! says Jack. Come right aboard and take the helm, I'm going ashore to see the old folks and I want my Flannel to be spotless. Between ourselves, we've been COALING—but not a word—war is war and I know you can keep a secret—Yes, it's dirty stuff is coal. What's happening in the North Sea? Well, there's still a lot of weather there—more than that I dare not tell you. But I can promise that when you know ALL we've done, our fame —but hard a'port, LUX, it's your fame I'm talking about—the fame of LUX for washing Flannels and Woollens. Why you make a lather like the wake of a fast cruiser, and as for the dirt, well—er—you just "STRAFE" DIRT and SAVE the fabric.

LUX WON'T SHRINK WOOLLENS.

In Packets everywhere, 1d., 2d., 3d. and 4d.

LEVER BROTHERS LIMITED, PORT SUNLIGHT.

Magazine advertisement, 1916

The CLEANEST fighter in the World— the British Tommy

The clean, chivalrous fighting instincts of our gallant soldiers reflect the ideals of our business life. The same characteristics which stamp the British Tommy as the *CLEANEST FIGHTER IN THE WORLD* have won equal repute for British Goods.

SUNLIGHT SOAP is typically British. It is acknowledged by experts to represent the highest standard of Soap Quality and Efficiency. Tommy welcomes it in the trenches just as you welcome it at home.

£1,000 GUARANTEE OF PURITY ON EVERY BAR.

The name Lever on Soap is a Guarantee of Purity and Excellence.

LEVER BROTHERS LIMITED, PORT SUNLIGHT.

Magazine advertisement, 1916

Paper bag dated 1917

Shaped display cards, 1953 and 1937

Showcard, *c.* 1915

The General

chewed a

FEEN-A-MINT
& constipation
was defeated

Feen-a-mint is the favourite family laxative; thorough and safe; delicious mint flavour. Get a packet from your chemist. For a free sample send your name and address and 1½d. in stamps (to cover postage), to Whites Laboratories Ltd., (Dept S.5), 52 Thames House, Westminster, S.W.1.

Magazine advertisement, 1936

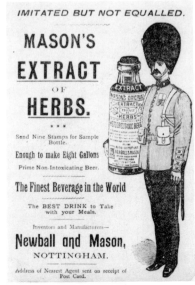

Magazine advertisement, 1916   Magazine advertisement, 1897   Magazine advertisement, 1936

Chocolate biscuit box, 1938

Matchbox, c. 1900

Pen nib box, c. 1915

Commemorative tin, 1902

Gas mantle box, c. 1910

Tobacco tin, c. 1900

Confectionery tin, c. 1915

Playing cards, c. 1915

Confectionery tin, c. 1935

Needle tin, c. 1905

Calendar card for 1910

Tacks tin, c. 1910

Razor blade display box, c. 1935

Paper novelty, c. 1895

70

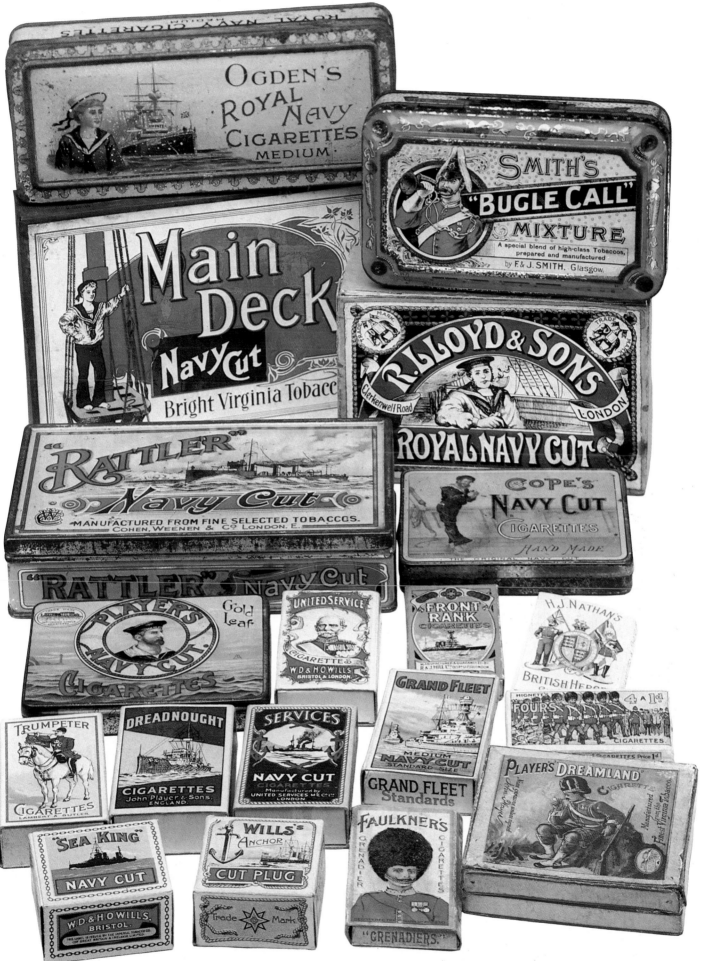

Tobacco and cigarette tins and packets from 1890 to 1920

JOHNNIE WALKER: "You've heard what the War Minister says about you! eh?"

AVIATOR (R.F.C.): "Yes! He describes us as your friends describe you."

JOHNNIE WALKER: "How's that?"

AVIATOR (R.F.C.): "The finest in the world."

JOHN WALKER & SONS, LTD., Scotch Whisky Distillers, KILMARNOCK.

Magazine advertisement, 1915

Magazine advertisement, 1918

The Royal Flying Corps was formed in 1912 with the Royal Air Force being created in 1918

Magazine advertisement, 1918

Showcard, c. 1940

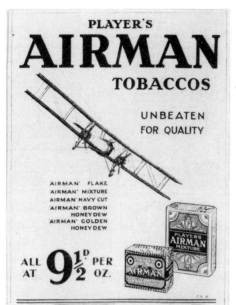

Magazine advertisement, 1931

Showcard, 1934

Magazine advertisement, 1932

Showcard, c. 1950

Showcard, c. 1940

GREAT BRITISH INSTITUTIONS

Two thousand years of justice, though often rough justice, has helped to mould many a great British institution. The country's legal system has been copied by many other democracies; the first organized police force was started in Britain, being founded by Sir Robert Peel in 1829; the country benefited from a respected educational system; and the Boy Scout movement was founded in 1908 by Lord Baden-Powell. The historic buildings of London, too, are an institution in their own right. All these, and more, found favour with manufacturers who wished to convey an impression of authority and reliability.

Showcard, *c.* 1925. Tower Bridge was opened in 1894

Trade card, *c.* 1900

Leaflet, *c.* 1905

Advertisement on the back of a promotional booklet, *c.* 1900

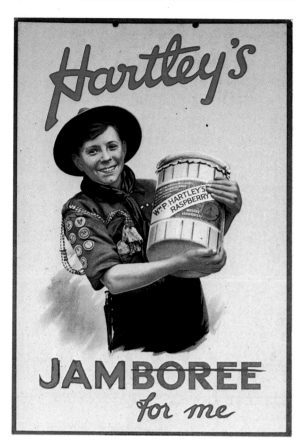

Showcard, *c.* 1925

Showcard, *c.* 1890. St Paul's Cathedral was rebuilt
following the Great Fire of 1666. Designed by Sir
Christopher Wren, the dome was completed in 1710

Showcard, 1912

Magazine advertisement, 1939

Give them a Trial

OGDEN'S
"GUINEA-GOLD"
CIGARETTES.

Magazine advertisement, 1897

Enamel sign, *c.* 1910

Tea wrapper, *c.* 1910

Blotting paper, *c.* 1930

Showcard, 1900

Matchbox label, *c.* 1925

Tobacco tin label, *c.* 1910

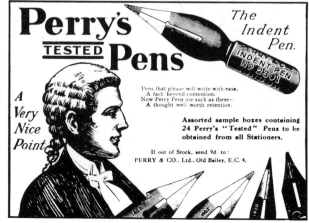

Magazine advertisement, 1918

Showcard, *c.* 1905

Magazine advertisement, 1937

Press advertisement, *c.* 1925

Magazine advertisement, 1897

Magazine advertisement, 1933

Magazine advertisement, 1898

Magazine advertisements, 1901

Magazine advertisement, 1901

Showcard, *c.* 1905

Showcard, *c.* 1895

Shaped trade card, *c.* 1910

Magazine insert, 1909

Shaped trade card, *c.* 1910

Magazine insert, 1911

Paper novelty, *c.* 1900

Both sides of a paper novelty, *c.* 1895

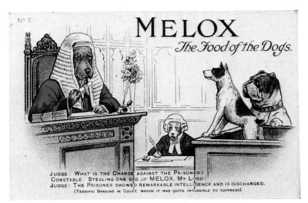

Post card dated 1922

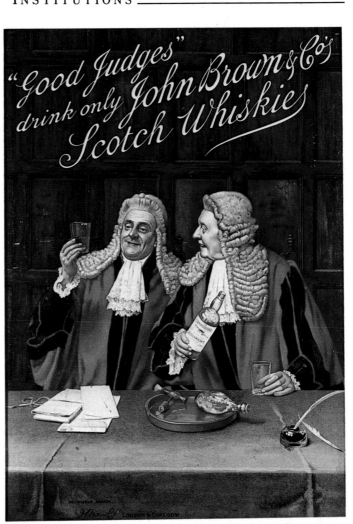

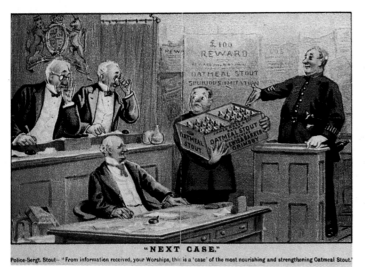

Leaflet, *c.* 1905

Showcard, *c.* 1890

Cough lozenge tin, *c.* 1900

Matchbox, 1905

Cigarette tin, *c.* 1895

Cough lozenge tin, *c.* 1910

Magazine insert, 1914

79

Showcard, c.1930

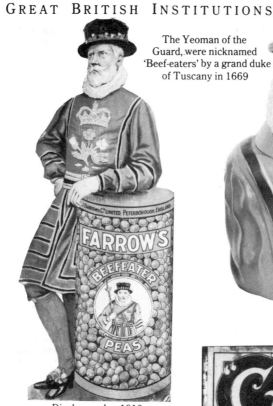

Display card, c. 1910

The Yeoman of the Guard, were nicknamed 'Beef-eaters' by a grand duke of Tuscany in 1669

Ice bucket, c. 1960

Snap fastener card, c. 1920

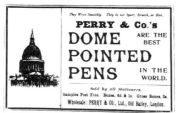

Magazine advertisement, 1897

Toffee tin, c. 1920

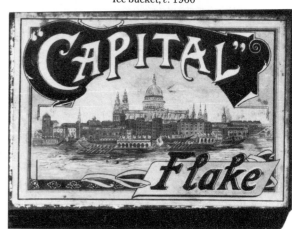

Tobacco tin, c. 1895

Razor blade wrapper
dated 1948

Detail from
poster, c. 1935

Matchbox label, c. 1920

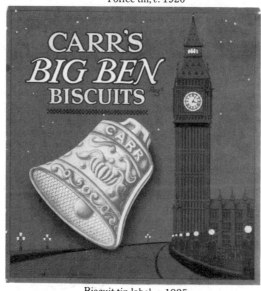

Biscuit tin label, c. 1925
The bell for Big Ben was cast in 1858

Anchovy paste jar, c. 1880

Toothpaste pot, c. 1900

Cigarette packet, *c.* 1910

Showcard, *c.* 1920

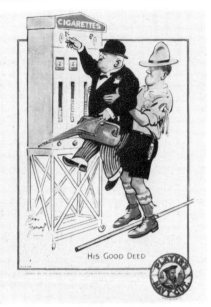

Magazine advertisement, 1927

Magazine insert, *c.* 1890

Shaped trade card, *c.* 1910

Magazine advertisement, 1905

Folding novelty, *c.* 1895

Leaflet, *c.* 1900

Magazine advertisement, 1889

UPSTAIRS, DOWNSTAIRS

The daily routine for domestic servants revolved around cleaning the house, washing and starching clothes, preparing meals and waiting for calls from 'upstairs'. The products used in these daily chores were continually developed so that tasks could be done more and more quickly, pleasantly and effectively, until eventually most households were able to manage without servants.

The images of 'upstairs, downstairs' scenes gave advertisers ample opportunity to demonstrate their wares.

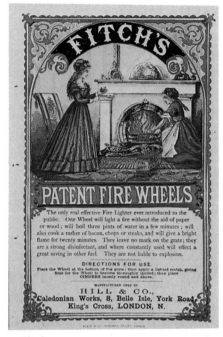

Magazine insert, *c.* 1870

Magazine insert, 1870

Trade card, *c.* 1885

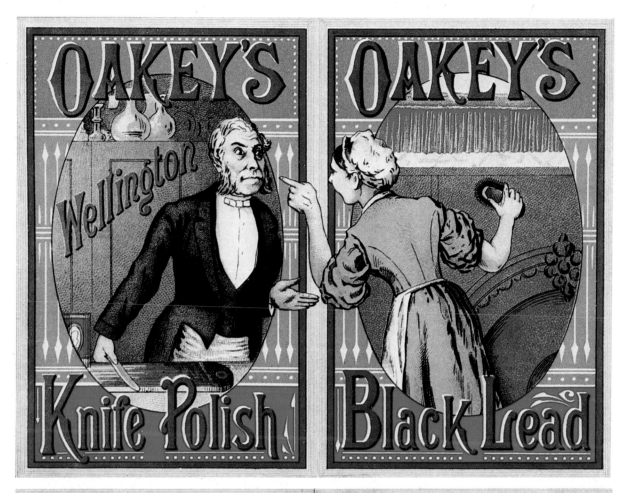

Folding leaflet, *c.* 1890

Magazine insert, *c.* 1890

Magazine insert, *c.* 1895

Magazine insert, *c.* 1900

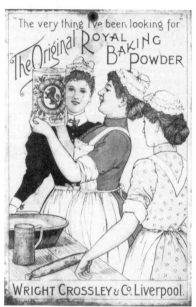

Showcard, *c.* 1895

Magazine insert, 1896

Leaflet, *c.* 1900

Magazine advertisement, 1897

Magazine advertisement, 1893

Magazine insert, *c.* 1895

Leaflet, 1903

Magazine insert, *c.* 1905

Magazine advertisement, 1888

Magazine advertisement, 1898

Magazine insert, 1912

Showcard, *c.* 1920

Magazine insert, *c.* 1900

Magazine insert, *c.* 1900

Showcard, 1898

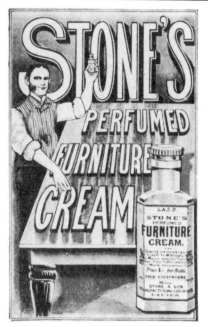

Magazine advertisement, 1910

Paper novelty, *c.* 1900

Shaped trade card, *c.* 1900

Magazine advertisement, 1911

Trade card, *c.* 1910

Matchbox label, *c.* 1920

Magazine advertisement, *c.* 1900

Magazine advertisement, 1910

Music sheet front cover, c. 1905

Showcard, 1903

Showcard, *c.* 1910

Movable paper novelty, *c.* 1910

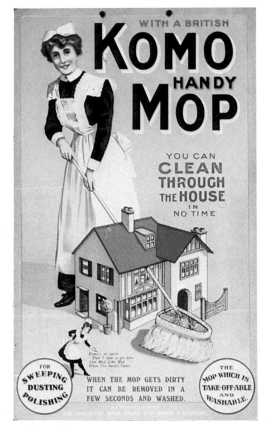

Showcard, *c.* 1905

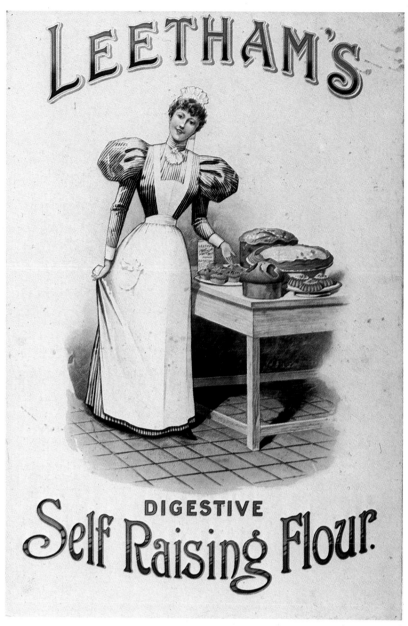

Showcard, *c.* 1890

Promotional booklet, 1897 (back cover)

Magazine insert, *c.* 1905

Leaflet, *c.* 1910

Magazine advertisement, 1917

Magazine advertisement, 1913

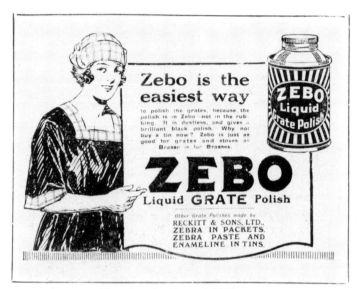

Magazine advertisement, 1922

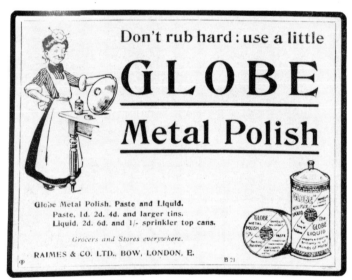

Magazine advertisement, 1910

Magazine advertisement, 1914

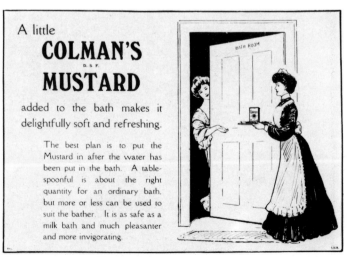

Magazine advertisement, 1906

Showcard, 1909

93

Trade card, *c.* 1890

Postcard, *c.* 1900

Shaped trade card, *c.* 1900

OVER 50 GOLD & PRIZE MEDALS AWARDED

Showcard, *c.* 1895

Trade card, *c.* 1895

Paper folding novelty, *c.* 1895

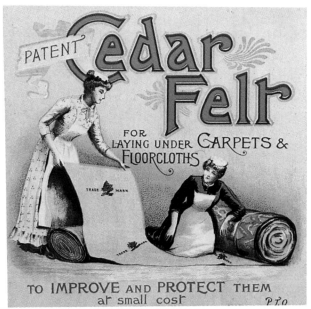

Leaflet, c. 1895

Paper folding novelty, c. 1895

BIRD'S CUSTARD POWDER
for your Custards.

BIRD'S BLANC-MANGE POWDER
for your Blanc-manges.

BIRD'S CRYSTAL JELLY POWDER
for your Jellies.

BIRD'S EGG POWDER
for your Cakes, Buns. &c

BIRD'S BAKING POWDER
for your Bread, Pastry, Puddings &c

Paper folding novelty, c. 1900

Paper folding novelty, c. 1895

"His Master's Voice"
or why the dinner was late.

Poster, c. 1905

Poster, *c.* 1925

Magazine advertisement, 1917

Magazine advertisement, 1920

Magazine advertisement, 1917

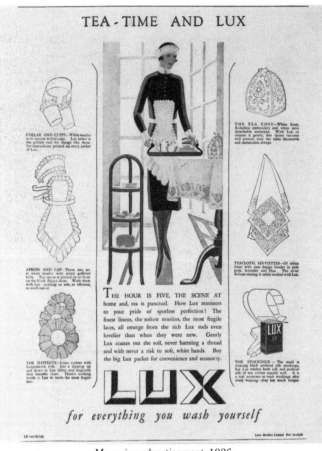

Magazine advertisement, 1926

Poster, *c.* 1920

Showcard, *c.* 1935

Trade magazine advertisement, 1920

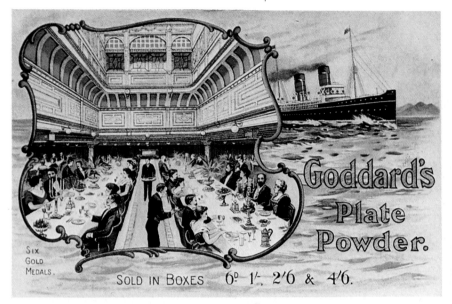

Leaflet, *c.* 1905

Magazine insert, *c.* 1900

Leaflet, *c.* 1905

Showcard, 1913

Postcard, *c.* 1905

Leaflet, *c.* 1900

Paper folding novelty, *c.* 1890

Display card, *c.* 1930

Showcard, *c.* 1925

TO PREVENT INFLUENZA, COLDS, CHILLS, AND WINTER ILLS

USE **BOVRIL** FREELY.

Magazine advertisement, 1893

Leaflet, *c.* 1905

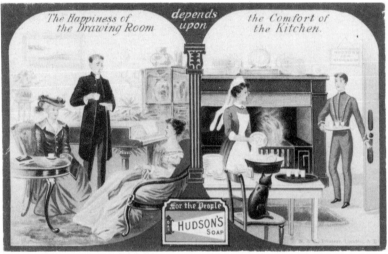

Trade card, *c.* 1895

Trade card, *c.* 1895

Magazine insert, *c.* 1895

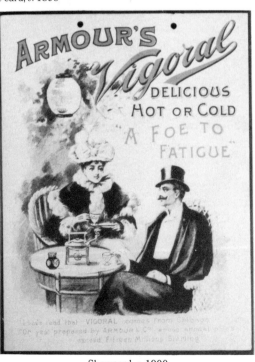

Showcard, *c.* 1900

Magazine advertisement, 1895

Society's palpably hit,
With a craze for the Card Game called PIT.
Ping Pong is played out,
So we all laugh and shout,
For there's Fun and Excitement in PIT,
    PIT is IT.
There is Fun and Excitement in PIT.

The "DAILY MAIL" says : "Pit is the latest game which Society has taken to
    its heart."
The "WORLD" says : "No one who has not played it can have any idea
    what a tonic it is for the spirits, and what peals of laughter it provokes."
The "GENTLEWOMAN" says : "Pit is now played everywhere."

*For sale by all first-class Dealers in Sports and Games, and at all
W. H. Smith and Son's Bookstalls.*

Price 2s. ; Gold Edge Edition, 3s.

PUBLISHED BY

PARKER BROTHERS, Lovell's Court, London, E.C.

Magazine advertisement, 1904

Showcard, 1909

Mr GOLD AND Mr FLAKE

"We're Mr. Gold and Mr. Flake,
Famed for the Cigarettes we make,
And every record we're out to break,
We do whatever we undertake :
We're men with WILLS of our own."

Showcard, 1929

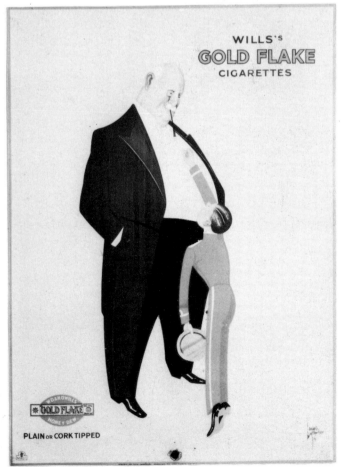

Showcard, 1935

Being a nation of explorers and colonists, Britons have carried with them their customs, traditions, pastimes and domestic comforts. Thus it was that British products were dispatched to all corners of the earth, whether biscuits to Africa or baby food to Australia. As British military forces moved around the world, their supplies followed, and doubtless this encouraged overseas awareness of Britain's products. Explorers, such as H. M. Stanley, who captured the public imagination found themselves depicted in advertisements recommending sustaining beverages, as did explorers in arctic regions.

Today, British communities abroad remain more determinedly British than Britons back home – the ritual of tea-time, for instance, being strictly observed – and British brands continue to be exported in their original livery long after their image has been updated in Britain.

Leaflet, *c*. 1895

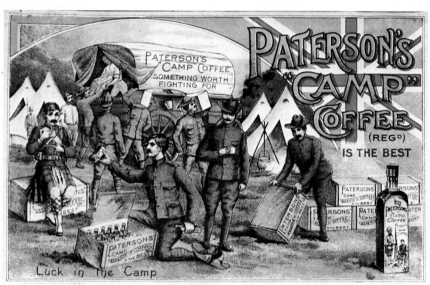

Leaflet, *c*. 1890

Boot lace box, *c*. 1890

Trade cards for Huntley and Palmers biscuits, *c*. 1890

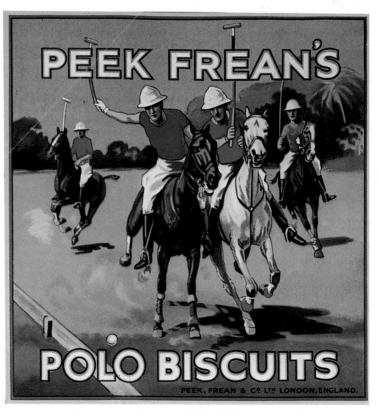

Poster, 1894

Matchbox label, *c.* 1910

Biscuit tin label, *c.* 1925

London Offices: 4, GREAT TOWER ST., LONDON, E.C.

Magazine advertisement, 1893

"HEALTH" COCOA ABSOLUTELY PURE

A QUARTER POUND TIN. COSTING 8ᴰ MAKES OVER 30 CUPS OF DELICIOUS COCOA

Leaflet, c. 1895

From POLE TO POLE

Cadbury's Cocoa and Chocolate

nourishing and sustaining in all climates and under all conditions.

NANSEN took a large quantity of CADBURY'S COCOA on his expedition towards the North Pole. Farthest North, 1896-1896.

SCOTT took 3500 lbs. of CADBURY'S COCOA and CHOCOLATE on his journey in the Antarctic regions. The Cocoa was his favourite beverage at lunch.

The Voyage of the "Discovery," 1901-1904

Press advertisement, 1918

TRIBUTE FROM THE ARCTIC · ·

"*...A year's supply of Barneys is among my personal stuff...*"

" I was most delighted to hear that my little letter of appreciation★ of Barneys impressed you so much as to be made the subject of one of your advertisements.

" The supply ship has only just arrived, bringing the yearly supplies and trade goods, etc. A year's supply of Barneys is also among my personal stuff. Last year the ship did not get in at all owing to very bad ice conditions, but it did not worry me, as, being an old hand at this game, I always make sure to have a two years' supply of every-thing on hand, of course, Barneys included."

★ *In this he said: " I have yet to find a Tobacco that can beat Barneys . . . Pleasures are few up here in the Far North, and for me at least smoking ranks number one, and only the best of tobaccos is good enough."*

The writer of this letter must be, without any doubt, the remotest Barneys smoker. He is stationed 400 miles within the Arctic Circle . . . cut off on occasion for two years from intercourse with the outer world. His letter did not reach us until a year after it was written.

It is no idle phrase when they say Barneys is a man's Tobacco, oftimes his friend. Think of this smoker, with probably the loneliest job on Earth . . . where a pipe of Barneys becomes the greatest pleasure life has to offer. " Only the best of tobaccos is good enough " in circum-

stances like his, and we are proud to think that Barneys Tobacco measures up to his conception of " the best."

It is given to few to be cast in the heroic mould, but the Barneys he praises, this " friendliest of all Tobaccos," is available to all. Maybe you would do well to try Barneys; there are three strengths: Barneys medium; Punchbowle full strength; Parsons Pleasure, mild, packed in the "EverFresh" Tin which, of a certainty, delivers Factory-freshness to smokers everywhere: 1s. 2d. Available also in "Ready-Fills" for quick, easy pipe-filling. Cartons of 12, 1s. 2d.

BARNEYS TOBACCO

Here is a Cigarette as good as Barneys Tobacco !

. . a high-class Virginia . . . in the Barneys tradition, costing only 6d. for 10. Barneys Cigarettes have a distinctive flavour and aroma that come from matured Virginia leaf, carefully selected and blended. The smoker of keen appreciation will find Barneys Ideal Cigarettes to be very good.

T.146) Made by John Sinclair Ltd., Bath Lane Factory, Newcastle-on-Tyne, England.

Magazine advertisement, 1939

Sir Henry Morton Stanley was born in Wales in 1841. He sailed to New Orleans at the age of 15 to seek out a new life. After many sea and war adventures, he was commissioned by the New York Herald to find out if the explorer, David Livingstone was still alive in Africa.

Stanley found him in 1871, greeting him with the legendary "Dr Livingstone , I presume". Stanley returned to Africa twice, making many geographical discoveries including tracing the course of the Congo. Though he had taken American citizenship in 1886, he became a British subject again in 1892. His book 'In Darkest Africa' was published in 1890.

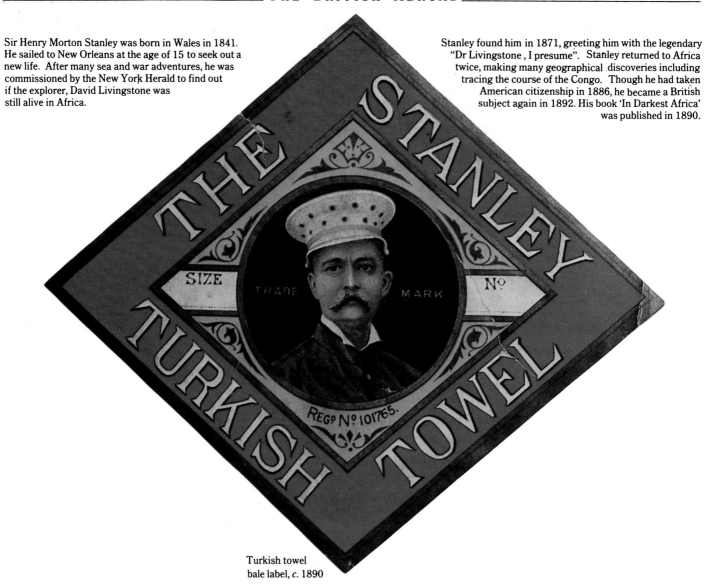

Turkish towel
bale label, *c.* 1890

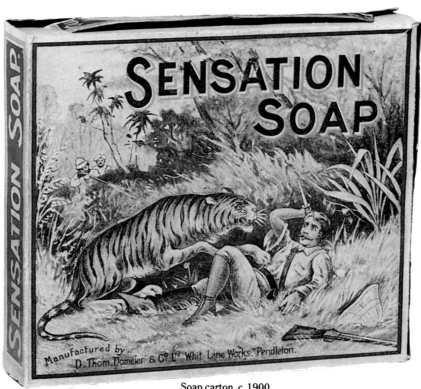

Soap carton, *c.* 1900

106

Showcard, 1910

Trade card, c. 1890

Magazine advertisement,
1888

Magazine insert, c. 1900

Magazine advertisement, 1898

Gold was discovered in 1896 in Klondike,
Canada. Within four years 30,000 people
had gathered around Bonanza Creek

Showcard, c. 1920

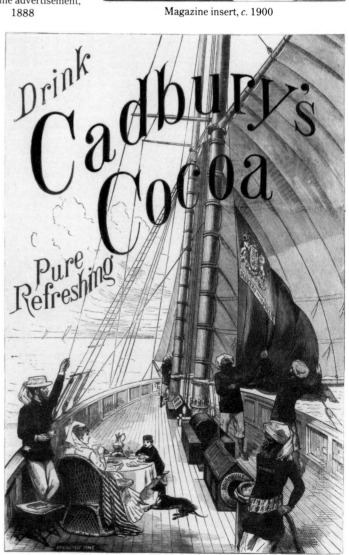

Magazine advertisement, 1884

Magazine advertisement, 1913

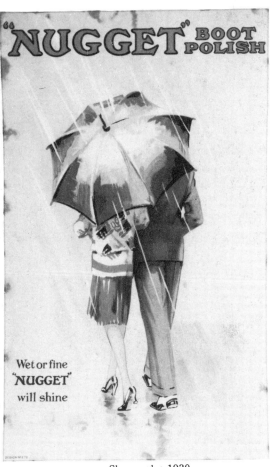

Showcard, c. 1930

Look Out FOR the Spout

but if by accident or otherwise you get wet, catch cold, or have any trouble with your lungs, **Look Out** for the nearest chemist who, for 1 1½ will sell you a tube containing six dozen of Geraudel's Pastilles — the best remedy for Coughs, Cold, Hoarseness, Sore Throat, &c., &c.

Magazine advertisement, 1897

Magazine advertisement, 1938

Showcard, c. 1935

When the weather turns to snow, the great national pursuit is throwing snowballs; a commentator in 1614, one of the coldest years, said: 'It's a sign that tradesmen and handy-craft have either little to doe, or else can doe little, by reason of the weather, when they throw by their tooles and fall to flinging of snow-bals'. Little it seems, has changed since then, except that for the past hundred years Camp coffee has been available to restore circulation.

Many of today's international sports originated in Britain, where native talent for organization and the creation of rules turned them from informal pastimes into competitive games. Cricket and football had their origins in the Middle Ages; but it was not until 1787 that the Marylebone Cricket Club came into being and formulated a set of rules. The Football Association was founded in 1863; the first British Open golf championship was played in 1860; and lawn tennis was formalized in 1874 and rapidly overtook croquet as the most popular social game.

Rules enabled teams to compete fairly with each other and, even though other countries took up these sports and have beaten the British at their own game, manufacturers still find the association of these images worthwhile.

Trade cards from two different series, *c.* 1885 and *c.* 1890

Showcard, c. 1890

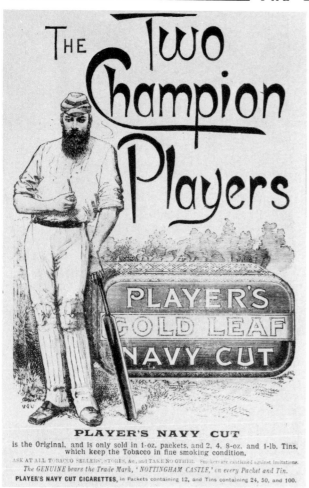

Magazine advertisement, *c.* 1890

Born in 1848, W. G. Grace (see left and page 115) was England's most renowned cricketer. He played first-class cricket from 1865 till 1900, hitting 121 centuries. A great athlete, on one occasion in 1866 he scored 224 runs not out for England v. Surrey and two days later won a race at Crystal Palace

Magazine advertisement, 1893

Magazine advertisement, 1897

Showcard, *c.* 1910

'Still scoring heavily!'

3,000,000 a week & Not Out.

Ogden's 'Guinea-Gold' Cigarettes

Three Pence per Packet of Ten.

Magazine advertisement, 1897

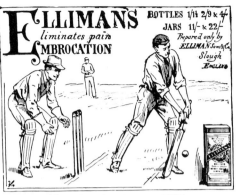

ELLIMANS liminates pain EMBROCATION

BOTTLES 1/1½ 2/9 & 4/-
JARS 11/- & 22/-
Prepared only by ELLIMAN Sons & Co Slough England

Magazine advertisement, 1897

MADE IN ENGLAND
THE CRICKET MATCH

Matchbox label, c. 1925

'IT'S AS PURE AS IT LOOKS!'

MEN are talking about Williams—the new Shaving Cream. It's as pure as it looks; pure white. No colouring matter; no irritants. Its cool, rich, quick lather makes for 25% easier, slicker razor-work. It's mild with the mildness of absolute purity. Ninety years' experience behind every tube! Naturally men are talking about Williams: the new Shaving Cream.

★FREE A large sample tube of Williams Shaving Cream : enough for 50 shaves. Send your name and address and 2d. stamp to cover postage to J. B. Williams Co. (Dept. B4), 40, Union Street, London, S.E.1.

Williams 1'6 SHAVING CREAM

also use WILLIAMS AQUA VELVA the AFTER-SHAVING preparation with the pleasant thrill. A splash of Aqua Velva invigorates the skin : keeps the face, all day, as Williams lather leaves it—satin-smooth and well-conditioned. 6d. 1'6. 2'6.

Magazine advertisement, 1931

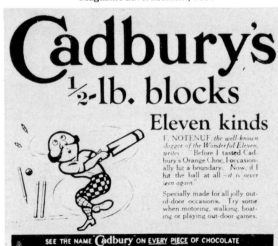

Cadbury's ½-lb. blocks

Eleven kinds

F. NOTENUF, the well-known slogger of the Wonderful Eleven, writes : "Before I tasted Cadbury's Orange Choc, I occasionally hit a boundary. Now, if I hit the ball at all—it is never seen again."

Specially made for all jolly out-of-door occasions. Try some when motoring, walking, boating or playing out-door games.

SEE THE NAME Cadbury ON EVERY PIECE OF CHOCOLATE

Press advertisement, 1928

Calendar for 1889

Showcard, 1903

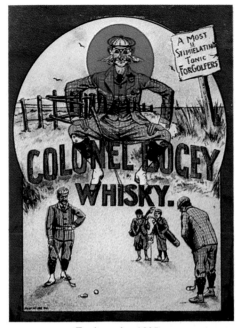

Trade card, c. 1895

Biscuit label, c. 1920

Hair pin box, c. 1910

Cigarette packet with tobacco tin and packet, c. 1925

Showcard, 1924

119

Magazine advertisement, 1897

Showcard, 1899

Magazine advertisement, 1927

Confectionery box, c. 1925

By the end of the nineteenth century women were
playing an active part in sports, particularly golf
and tennis, and advertisements delighted in the
use of the female form swinging a golf club or
tennis racket. It was a time of women's
emancipation, and competition was not just left
to the men. The first women's tennis tournament
was played in 1879, in Dublin. The ladies' Golf
Union was formed in 1893, and it was they who
created a universally acceptable form
of handicapping.

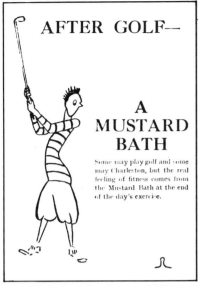

AFTER GOLF—

A
MUSTARD
BATH

Some may play golf and some may Charleston, but the real feeling of fitness comes from the Mustard Bath at the end of the day's exercise.

Magazine advertisement, 1928

"HIT IT MAN!"

Please use this blotter.

P. SMITH & SON,
(STUART W. SMITH.)
Builders and Decorators,
GLOUCESTER STREET, CLIFTON.

Every Description of House Repairs from Roof to Foundation.
Send to us for Quotations. Tel. No. 2272.

Blotter card, *c.* 1930

When his ball is on the green
And he must putt it true & clean
Says the golfer

Sharp's the word
and
Sharp's the Toffee
I like best of all

Magazine advertisement, 1933

IF YOUR THROAT IS SORE!

AND YOU CAN'T CALL "FORE"!

TRY

KILKOF KUBES

KILKOF KUBES
TACKLE THAT TICKLE
PARKINSONS Lᵗ

2D.
PER OUNCE

THEY'RE SURE TO SCORE.

PARKINSON'S LTD BURNLEY.

Press advertisement, *c.* 1925

Since when have YOU been using GIBBS!

Since I had a hole in one!

Don't wait for a minor calamity to your teeth before you enjoy the benefit of Gibbs Dentifrice. Your teeth need the unceasing care of Gibbs NOW. Gibbs Dentifrice kills germs, removes everything which might cause decay, cleans and polishes your teeth to gleaming whiteness. Its fragrant, antiseptic foam neutralises acids, makes your gums firm, your whole mouth feel delightfully toned up and refreshed. Do as Dentists do and advise—use Gibbs Dentifrice twice daily. Don't deny yourself of its benefits a moment longer.
CHANGE TO GIBBS TODAY.

Your teeth are Ivory Castles—defend them with

Gibbs DENTIFRICE

Solid
In Tins,
7½d. 1 -, 1 6
Refills 11d.

IN YOUR FAVOURITE FORM

Paste
In Tubes,
6d and 1 -

D. & W. GIBBS LTD., LONDON, E.1.

Magazine advertisement, 1938

Showcard, c. 1925

Shaped display card, 1928

Showcard, c. 1930

Two layer display card, c. 1930

The Dream of the Golfer who forgot his Guinness a day

Caddie : " Your Tee is ready, sir "
Absent-minded Golfer : " No thanks, I'll have a Guinness."

" Have a glass of Guinness when you're Tired."

" What should I take here, Caddie ? "
" I should take a Guinness, sir ! "

Cartoon advertisements from a promotional booklet of 1937

Showcard, *c.* 1930

Magazine advertisement, 1895

Magazine insert, *c.* 1900

Leaflet, *c.* 1895

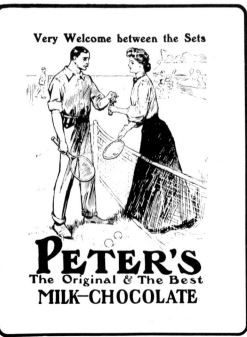

Magazine advertisement, 1906

Magazine advertisement, 1925

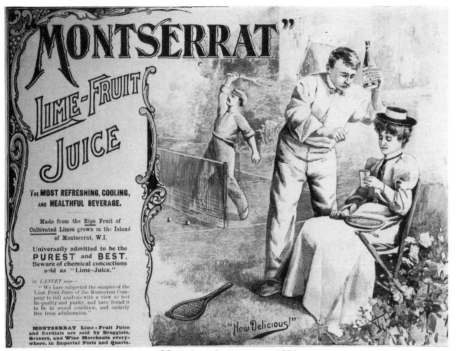

Magazine advertisement, 1898

Magazine advertisement, 1932

Press advertisement, 1928

Magazine advertisement, 1922

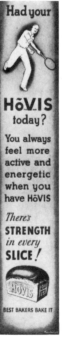

Magazine advertisement, 1936

Magazine advertisement, 1939

Press advertisement, c. 1930

Biscuit label, c. 1935

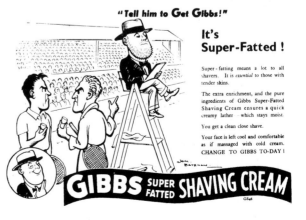

Magazine advertisement, 1936

125

Showcard, *c.* 1880

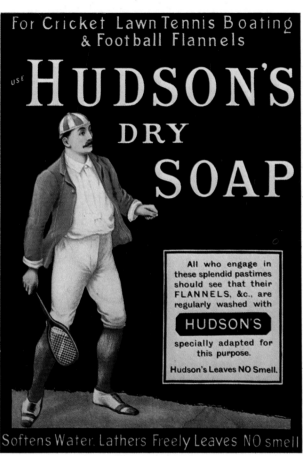

Showcard, *c.* 1885

Trade card, *c.* 1890

Showcard, *c.* 1900

Showcard, c. 1935

Leaflet, 1877

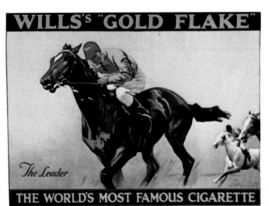

Poster, 1928

Needle tin, c. 1905

Poster, 1925

Artwork for toffee tin, c. 1930

Custard powder tin, c. 1925

127

Showcard, *c.* 1900

Confectionery box label, *c.* 1890

Biscuit tin label, *c.* 1925

Magazine advertisement, 1926

Cigarette tin label, *c.* 1925

Whisky label, *c.* 1910

Magazine insert, *c.* 1895

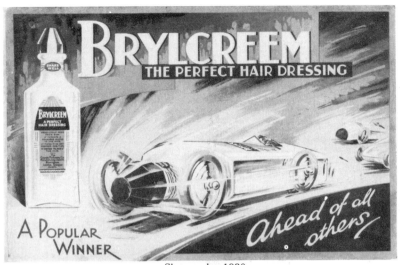

Showcard, *c.* 1930

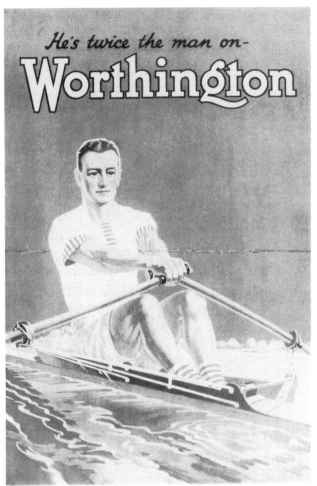

Magazine advertisement, 1928

Poster, *c.* 1935

Showcard, 1938

Magazine advertisement, 1900

MAYPOLE SOAP.

DYES ALL COLOURS

Bookmark, *c.* 1905

MURRAY'S

Varsity Mixture

R.S. MURRAY & C° L^{TD}. LONDON. E.C.

Confectionery box lid, *c.* 1910. The first Oxford v. Cambridge
boat race was held in 1829

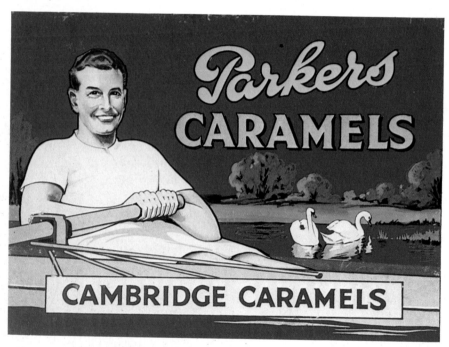

Parkers CARAMELS

CAMBRIDGE CARAMELS

Artwork for inside of lid of confectionery tin , *c.* 1925

Cigarette packet, *c.* 1890
depicting a Cambridge
University oarsman smoking
a cigarette that 'contains
less nicotine'

PLAYER'S NAVY CUT
Tobacco & Cigarettes

The Nation's Choice

Poster, 1926

PLAY UP LADS FOR THE
FOOTBALL SAUCE
A PERFECT RELISH
for Fish, Hot and Cold Joints,
Chops, Steaks, Soups, etc.
A. HARROP & SONS
SHEFFIELD

Sauce label, *c.* 1910

MEDIUM
GOLD PLATE
CIGARETTES.
GALLAHER L^{TD}
THE INDEPENDENT FIRM

Cigarette packet, *c.* 1910

The playing of bowls goes back to ancient times, and Sir Francis Drake is said to have been playing the game when the Spanish Armada was sighted in 1588. Rules were not standardized until 1849

Showcard, 1932

Cigarette tin, *c.* 1895

Cigarette packet, *c.* 1900

Detail from poster, *c.* 1905

Matchbox label, 1891

Chocolate box, *c.* 1930

Showcard, *c.* 1925

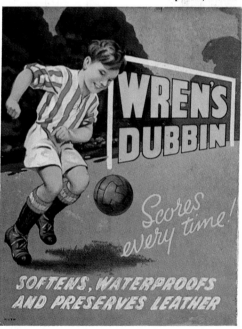

Showcard, *c.* 1950

131

"Strength and Staying Power."

CADBURY'S COCOA

IS ABSOLUTELY PURE

CADBURY'S COCOA

SUSTAINS
  AGAINST FATIGUE.
INCREASES MUSCULAR
  STRENGTH.
GIVES PHYSICAL ENDURANCE
  AND STAYING POWER.

CADBURY'S ABSOLUTELY PURE COCOA is a refined concentration of the strength-sustaining and flesh-forming constituents of the Cocoa Nib. Delicious, nutritious, easily digested, and of great economy, a Sixpenny Packet yielding fourteen large breakfast cups of perfect Cocoa.

Magazine advertisement, 1888

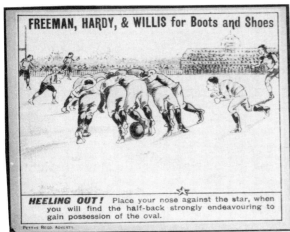

FREEMAN, HARDY, & WILLIS for Boots and Shoes

HEELING OUT! Place your nose against the star, when you will find the half-back strongly endeavouring to gain possession of the oval.

Puzzle card, *c.* 1905

CADBURY'S COCOA

ABSOLUTELY·PURE THEREFORE BEST
REFRESHING·SUSTAINING·INVIGORATING

Magazine advertisement, 1901

Stout bottle label, *c.* 1930

The English public schools and universities of the early nineteenth century were the main exponents of football, but each had their own rules.
By the end of the 1860s, a common set of rules had been devised and that led to teams being able to compete on equal terms, and thus the popularity of the game spread. Rugby football, as its name denotes, was developed at Rugby School in the 1830s, the Rugby Football Association being formed in 1871

OUTDOORS as indoors Tootal Piqué combines effectiveness with comfort and economy. It possesses that most essential virtue to-day—endurance, despite frequent visits to the wash.

TOOTAL PIQUÉ

This popular Tootal line is double-width; it is specially strengthened between the cords to prevent splitting, yet it never goes harsh or stiff, if carefully laundered. In great demand for washwear of every kind, especially for the cool, serviceable frocks and overalls needed by ladies engaged in hospital and other war-work. Always see name Tootal Piqué on Selvedge.

Four widths of cords and fancy patterns.
2/2 the double-width yard (43-44 inches), at drapers everywhere. Patterns free from TOOTALS, Dept. B 4, 132, Cheapside, London, E.C.

Magazine advertisement, 1916

Modern croquet dates back to 1857 when John Jaques wrote a book on the game and also manufactured the sets. The first championship was played at Evesham, Worcestershire, in 1867

SAVED.
SO MAY YOUR SIGHT BE, IF YOU TAKE OPTICAL ADVICE IN TIME

Showcard, *c.* 1935

Showcard, 1948

TAKE MY "CUE" IF THE COLD YOU DREAD

EAT KILKOF KUBES WHEN GOING TO BED.

2D. PER OUNCE

PARKINSONS LTD. BURNLEY.

Press advertisement, *c.*1925

Leisure pursuits come for all seasons: tobogganing and party-going in the winter, punting and cycling in the summer. The pre-television era allowed time for hiking; maypole dancing was the exercise approved by the Aesthetic Movement. Such energetic and graceful activities provided the ambience for innumerable advertisements and food wrappers.

Poster, 1877

Showcard, *c.* 1895

Poster, *c.* 1885

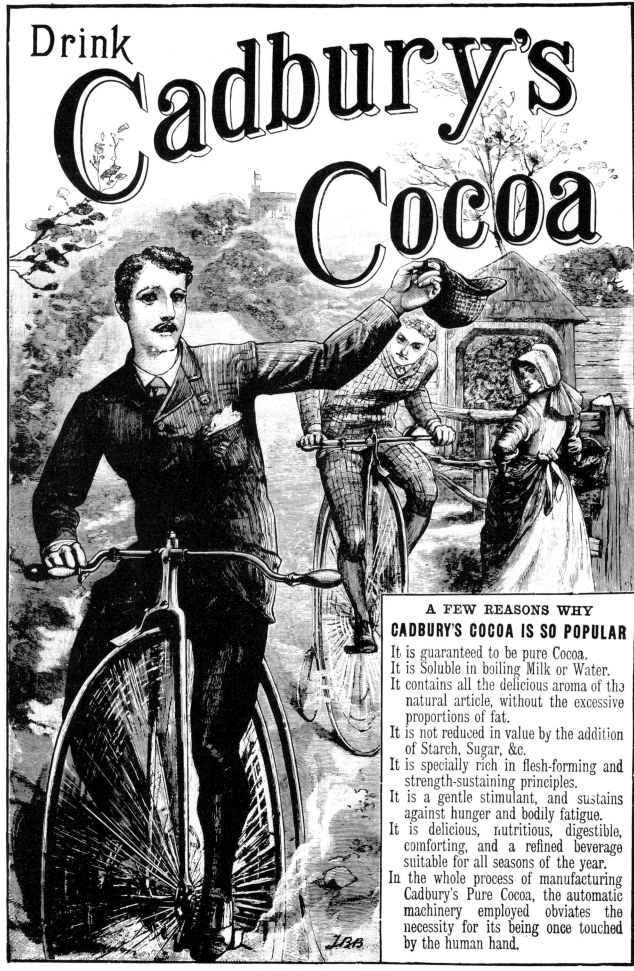

Drink Cadbury's Cocoa

A FEW REASONS WHY
CADBURY'S COCOA IS SO POPULAR

It is guaranteed to be pure Cocoa.

It is Soluble in boiling Milk or Water.

It contains all the delicious aroma of the natural article, without the excessive proportions of fat.

It is not reduced in value by the addition of Starch, Sugar, &c.

It is specially rich in flesh-forming and strength-sustaining principles.

It is a gentle stimulant, and sustains against hunger and bodily fatigue.

It is delicious, nutritious, digestible, comforting, and a refined beverage suitable for all seasons of the year.

In the whole process of manufacturing Cadbury's Pure Cocoa, the automatic machinery employed obviates the necessity for its being once touched by the human hand.

Magazine advertisement, 1887

Magazine insert, *c.* 1905

Magazine insert, *c.* 1900

Magazine advertisement, 1895

Biscuit tin label, *c.* 1895

Cycling as a leisure-time pursuit benefited greatly in 1868 from the introduction of rubber tyres for wheels, instead of iron ones. The penny-farthing cycle was popular in the 1870s and 1880s. However, cycles with equal-sized wheels prevailed for safety reasons, and with the invention of pneumatic tyres by J. B. Dunlop in 1888 (see page 24), the popularity of cycling increased rapidly. Ladies welcomed the arrival of the drop-frame cycle, which accommodated the long skirts of the 1890s

Matchbox label, *c.* 1910

Leaflet, 1903

Magazine advertisement, 1906

137

SOUTHWELL'S LEMON SQUASH

REGISTERED
Chas. Southwell & Co
TRADE MARK

Showcard, *c.* 1895

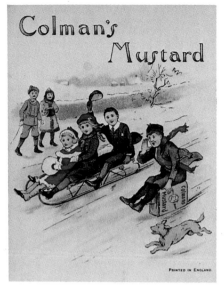

Promotional booklet, 1903 (back cover)

Poster, c. 1880

Paper novelty, c. 1900

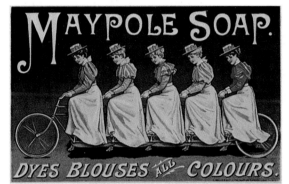

Magazine insert, c. 1900

Corset box, c. 1905

Confectionery tin, c. 1895

Magazine insert, c. 1900

Trade card, c. 1905

Paper bag, *c.* 1900

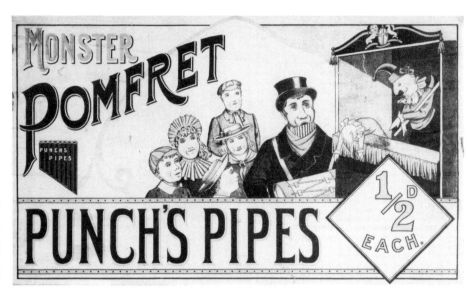

Label inside wooden display box, *c.* 1900

Samuel Pepys saw 'an Italian puppet play' being
performed in Covent Garden in 1662. The Punch and
Judy show continued to be a common sight in London
streets up to the Edwardian era

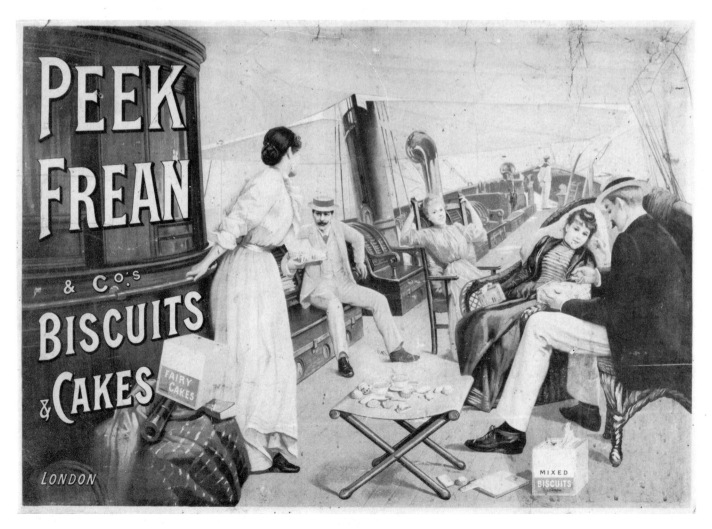

Showcard, *c.* 1890

FREEMAN, HARDY & WILLIS LIMITED

Puzzle Find
THE BOATMAN

Puzzle Find
THE HOST

FOR BOOTS AND SHOES.

Puzzle card, *c.* 1905

ROYAL DEVONSHIRE SERGE.

All the latest fashionable Colours, Mixtures, and Textures For Ladies', Children's, Gentlemen's and Boys' Dress. Hard Wear Guaranteed. Price from 1s. 6½d. per yard.

"SEA WATER CANNOT HURT IT."

On the authority of the *QUEEN*, the Dress Fabrics specially produced by Messrs. SPEARMAN and SPEARMAN stand unrivalled for Beauty, Durability, and General Usefulness.

Any Length Cut, and Carriage Paid to London, Dublin, and Glasgow.

CAN ONLY BE OBTAINED GENUINE OF THE SOLE FACTORS,

he won't spoil his things Auntie, they are made of SPEARMAN'S ROYAL DEVONSHIRE SERGE like yours, and mine.

SPEARMAN & SPEARMAN, (ONLY ADDRESS) PLYMOUTH.

Magazine advertisement, 1881

RECKITT'S Blue

Showcard, *c.* 1900

THE SUMMER GIRL

TAKES

BEECHAM'S PILLS

Magazine advertisement, 1920

URODONAL

RHEUMATISM AND SEA AIR

Rheumatism
Gout
Gravel
Arterio-Sclerosis,
Neuralgia,
Migraine,
Sciatica.

Is it advisable for rheumatic subjects to go to the seaside? It will perhaps be argued that those who were born at the seaside, or who have lived there the greater part of their life, are specially favoured on account of having become accustomed to the atmosphere, while tourists who only come for a few days, and are therefore strange to it, cannot claim the same privileges. That may be the case, but it still remains to know whether sea air itself is apt to aggravate rheumatic pains.

Precautions must, of course, be taken, and the best way of preventing attacks of rheumatism at the seaside or anywhere else is to neutralise the drawbacks caused by humidity and the risks of over-eating or other imprudences. The only thing to do is, therefore, to combat the over-production of uric acid by dissolving and eliminating it as fast as it is formed. Nothing can be easier than to do this with the help of URODONAL, which "dissolves uric acid as easily as hot water dissolves sugar." This auxiliary and harmless precaution is moreover necessary not only at the seaside, but should be adopted almost anywhere at this time of the year, when change of air, exposure, and outdoor life tend to stir up the blood.

"*I thought you were forbidden to bathe on account of your Rheumatism?*" "*Oh, but now I take URODONAL I no longer suffer from Rheumatism*"

DR. DAURIAN,
Paris Medical Faculty.

URODONAL, prices 1s. and 1½s. Prepared at Chatelain's Laboratories, Paris. Can be obtained from all chemists and drug stores, or direct, post free, from the British and Colonial Agents, HEPPELLS, Pharmacists and Foreign Chemists, 164, Piccadilly, London, W., from whom also can be had, post free, the full explanatory booklets, "Scientific Remedies," and "Treatise on Diet."

Agents in Canada: ROUGIER FRÈRES, 83, Rue Notre Dame Est, Montreal, Canada. *Agents in U.S.A.*: GEO. WALLAU, 1, 2, 6, Chef Street, New York, U.S.A. *Agent for Australia and New Zealand*: BASIL KING Malcolm, Malcolm Lane, Sydney. G.P.O. 2259b. *Sole Agents for India, Burma and Ceylon*: ATHERTON & CO., 8, Clive Street, Calcutta

Magazine advertisement, 1903

Poster, *c.* 1925

Confectionery tin, *c.* 1935

Toffee tin, *c.* 1925

Chocolate box, *c.* 1925

JAZZ DYE SOAP
IN FINE POWDER
Just the shade you like

Magazine insert, *c.* 1925

Dye soap packet, *c.* 1930

Jazz seems to have been born
in New Orleans, and by the
early 1920s had supplanted
Rag time as the popular music
of the day. Fancy-dress parties
were also the rage

Magazine advertisement, 1929

142

Showcard, *c.* 1925

Leaflet, *c.* 1895

Dye packet, *c.* 1900

Magazine advertisement, 1928

The plaited maypole, a Continental custom, was introduced into this country by John Ruskin in 1888

Novelty card, *c.* 1895

Trade card, *c.* 1905

Biscuit tin label, *c.* 1895

BEECHAM'S PILLS

ENSURE GOOD HEALTH.

Magazine advertisement, 1911

GOOD SPORT

under these circumstances is most enjoyable, with PETER'S to promote sociability and provide refreshment

PETER'S
The Original & The Best
MILK-CHOCOLATE

Magazine advertisement, 1906

The Man who COUGHED
as the Colonel Fought a Fish...

KENSITAS would
have saved him!

Be careful in your choice of cigarettes! Choose KENSITAS—the only cigarette in the United Kingdom that offers the throat protection of that exclusive Private Process which includes the use of modern Ultra-Violet Rays—the process that expels certain biting, harsh irritants naturally present in every tobacco leaf. KENSITAS offers the finest, choicest, real Virginia tobacco, plus throat protection. No wonder 1004 British Doctors have stated KENSITAS to be less irritating. No wonder KENSITAS are always kind to your throat.

"With **Ultra-Violet** Rays"
Your Throat Protection – against irritation – against cough

"As good as really good cigarettes can be

"Your
Kensitas
Cigarettes Sir"
REAL VIRGINIA

TWENTY FOR ONE SHILLING

Magazine advertisement, 1931

144

Showcard, c. 1910

Roller-skating, introduced in England from America in the early
1870s (see page 134), reached the height of its popularity
between 1906 and 1912 when rinks were opened all over the country

Magazine advertisement, 1890

Tobacco containers 1900 to 1925

Postcard, *c.* 1920

Biscuit label, *c.* 1925

Promotional booklets (back covers),
1897 and 1898

Trade card, *c.* 1895

Toffee tin, *c.* 1930

Showcard, *c.* 1920

Poster, 1927

146

Tin sign, *c.* 1930

Showcard, *c.* 1935

Showcard, *c.* 1930

Showcard, *c.* 1930

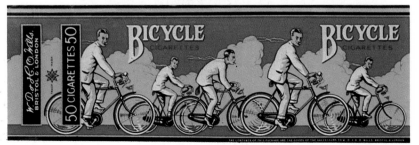

Cigarette tin label, *c.* 1930

Magazine advertisement, 1923

Chocolate biscuit box, *c.* 1935

Toffee tin in shape of ship's funnel, *c.* 1935

Magazine advertisement, 1933

Magazine advertisement, 1933

148

Magazine advertisement, *c.* 1930

A delightful tea
for all occasions—
BARBER'S TEA

Press advertisement, 1932

Price card, *c.* 1935

SUMMER SHELL PETROL
is seasonally blended to suit the prevailing temperature
and, by the way, to be up-to-date you must Shellubricate too

Magazine advertisement, 1931

Magazine advertisement, 1923

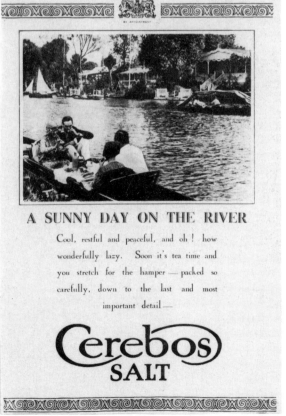

A SUNNY DAY ON THE RIVER

Cool, restful and peaceful, and oh! how
wonderfully lazy. Soon it's tea time and
you stretch for the hamper — packed so
carefully, down to the last and most
important detail —

Cerebos SALT

Magazine advertisement, 1928

149

The great British 'cuppa' has been the consolation of generations, whether tea, cocoa or coffee. It is such an important (if not ritualistic) part of the British way of life that disputes have arisen over the frequency or length of the tea or coffee break at the work place. Advertisers naturally tended to depict rather posh tea-time scenes, cosy cocoa-drinking sessions and picturesque picnics, rather than workmen imbibing strong drinks in pubs.

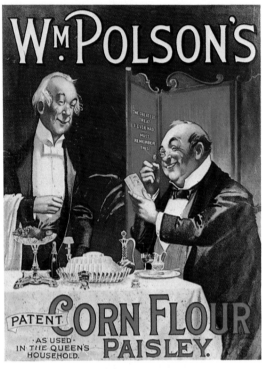

Showcard, *c.* 1895

Showcard, *c.* 1900

Showcard, *c.* 1900

Showcard, c. 1895

Leaflet, c. 1900

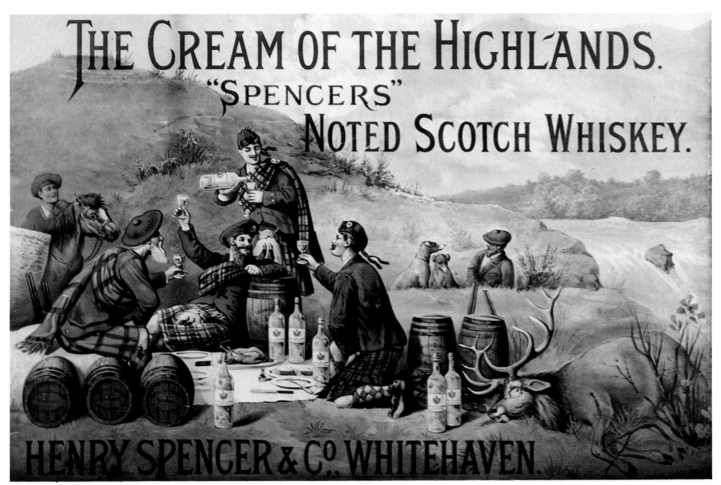

Showcard, c. 1895

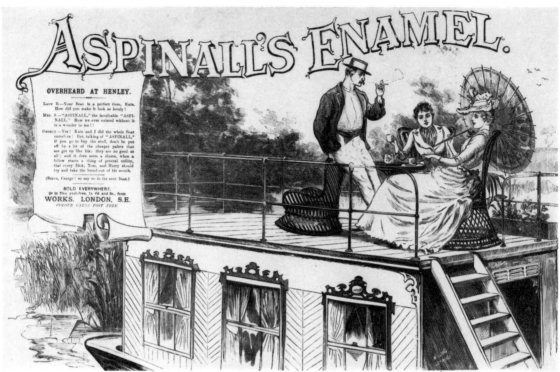

Magazine advertisement, 1890

Magazine advertisement, 1909

Magazine advertisement, *c.* 1905

Trade card, *c.* 1890

Trade card, *c.* 1895

Leaflet, *c.* 1900

In 1880 the Great Northern Railway provided dining cars for first class passengers, and this was soon followed by the Midland Railway who offered dining car facilities for everyone. However, it was the Great Western Railway who in 1890 put into service the carriage with a corridor, thus eliminating the special train stops at stations for passengers to take refreshments.

Showcard, *c.* 1895

Magazine advertisement, 1915

Showcard, *c.*1895

Leaflet, *c.* 1900

Showcard, *c.* 1900

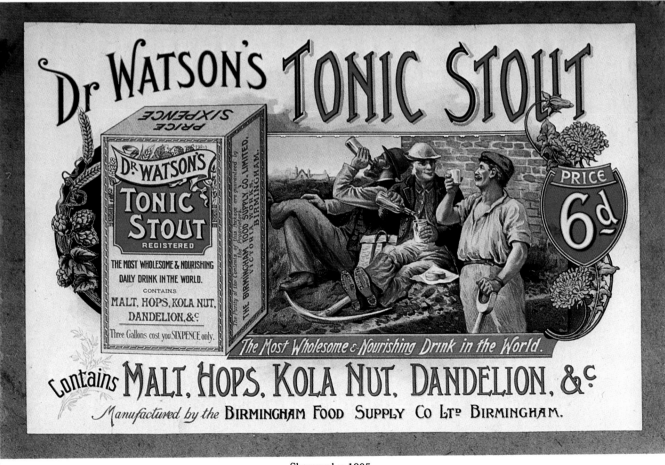

Showcard, *c.* 1905

"Mummy doesn't mind, 'cos it's Tobralco and washes so nice."

Tobralco is always in demand as a sound and beautiful wash fabric. Just now it is required specially for charming sports frocks and other outdoor dresses—nice to wear and easy to do up.

Tobralco is so serviceable and washes so beautifully that its dainty white fascinating colors and exclusive designs give pleasure through a long season of constant wear. No starch required when doing up. For yourself and children you will find the choice of Tobralco patterns this year more pleasing than ever—see the small Chintz designs, and smart Gingham designs in Tobralco **for** extra hard wear. Your favorite color is in the Tobralco range — *and guaranteed fast*

British-made·Cotton·Wash·Fabric

TOBRALCO
REG?

White & Wide Variety of Colors

The name Tobralco appears on every yard of Selvedge.

$9\frac{3}{4}$d.—a yard—Self-White. 27-28 inches wide | Guaranteed fast colors and Black— All same width. $10\frac{3}{4}$d.

See Tobralco at your drapers, or send for Free Patterns to Tobralco, Dept. 4, 132 Cheapside, London, E.C.

TOOTAL BROADHURST LEE CO. LTD., Manufacturers also of Tarantulle for dainty home-sewn lingerie; Tootal's Piqué, double width; Lissue Handkerchiefs for Ladies; Pyramid Handkerchiefs and Tootal Shirtings for Men.

14/26

48257

PAPER PATTERNS of these fashions specially recommended for making with Tobralco, can **be** obtained for 7d. each, post free, from WELDONS LTD., 31, Southampton Street, Strand, London, W.C.

48240

Magazine advertisement, 1914

Showcard, *c.* 1935

AT BREAKFAST TIME.

What a rush and a scramble to get through with the porridge when it is followed by

Laitova Lemon Cheese

The daily spread for the children's bread.

It's just the most delicious dainty you ever tasted. The kiddies can't have enough. It's wholesome and nourishing too, besides being so economical. Get a jar for breakfast to-morrow!

Laitova is a most welcome change from the usual bacon for the grown-ups.

It is now packed in dainty hygienic jars, and your grocer sells it in 7d. and other sizes.

SUTCLIFFE & BINGHAM. Ltd.. Cornbrook. MANCHESTER.

Magazine advertisement, 1916

Great success of the SHILLING HOLIDAY

CADBURY'S HOLIDAY

chocolate — with almonds and raisins

packed in a stiff cardboard carton. which keeps the chocolate clean & tidy.

Magazine advertisement, 1929

SUMMER TIME

is pic-nic time. How the youngsters enjoy these al-fresco outings and revel in the delights of woodland life! Then comes tea-time! The sticks to be gathered, the kettle to be boiled, the cloth to be laid, and finally the basket to be opened. With what a shout of welcome the youngsters greet their favourite—

The daily spread for the children's bread

It's really so delicious that the children are always looking forward to having it—you can't give them too much. Grown-ups like it just as much, and the housewife likes it too—*it saves the butter bill.*

Any day and every day Laitova will receive a hearty welcome, but more than ever on pic-nic days. Get an 8½d. jar from your grocer to-day.

SUTCLIFFE & BINGHAM, LTD., Cornbrook, MANCHESTER.

Magazine advertisement, 1916

Showcard, c. 1910

Magazine advertisement, 1920

Showcard, c. 1925

Biscuit tin label, *c.* 1930

Biscuit tin label, *c.* 1930

Sardine tin label, *c.* 1920

Showcard, *c.* 1930

159